John Harvie-Brown

The Capercaillie in Scotland

Elibron Classics
www.elibron.com

Elibron Classics series.

ISBN 1-4212-5175-2 (paperback)
ISBN 0-543-69038-5 (hardcover)

This Elibron Classics Replica Edition is an unabridged facsimile of the edition published in 1888 by David Douglas, Edinburgh.

THE CAPERCAILLIE IN SCOTLAND

WITH SOME ACCOUNT OF THE

EXTENSION OF ITS RANGE SINCE ITS
RESTORATION AT TAYMOUTH
IN 1837 AND 1838.

Glendochart

THE

CAPERCAILLIE IN SCOTLAND

WITH SOME ACCOUNT OF THE EXTENSION OF ITS RANGE SINCE ITS RESTORATION AT TAYMOUTH IN 1837 AND 1838

BY J. A. HARVIE-BROWN, F.Z.S.

MEMBER OF THE BRITISH ORNITHOLOGISTS' UNION, ETC.

"And from the pine's high top brought down
The Giant Grous while boastful he display'd
His breast of varying green, and crow'd and clapp'd
His glossy wings."

GISBORNE—*Walks in a Forest.*

EDINBURGH: DAVID DOUGLAS

MDCCCLXXXVIII

Dunipace House, *20th April 1879.*

My dear Professor Newton,

As you first drew my attention to the subject treated of in the following pages, and while my inquiries were proceeding, kindly assisted me by continued advice, I desire to inscribe to you this little volume, and to remain yours very truly and obliged,

THE AUTHOR.

To A. Newton, Esq., M.A., F.R.S.,
Professor of Zoology and Comparative Anatomy
in the University of Cambridge.

CONTENTS.

INTRODUCTION.

PART I.

PART II.

PART III.

PART IV. (*With a Map.*)

Increase and Extension of Range—*Continued.*

PART V.

PART VI.

APPENDIX.

INTRODUCTION.

In the autumn of 1877, Professor Newton of Cambridge intimated to me that he desired to have some account of the increase and extension of range of the Capercaillie in Scotland. Although possessing a fairly accurate conception of its general distribution, and the lines of its advance outwards from Taymouth, where it was restored in 1837-8, I found that I was wanting in the more minute details which it would be necessary for me to possess before I could furnish a suitable reply. Accordingly, I began inquiries; at first simply with a view to furnishing Professor Newton with a short summary for his new edition of Yarrell's "*British Birds.*" But information of such valuable, suggestive, and interesting kind came to hand, that I soon conceived the project of treating the subject more exhaustively. I reflected, moreover, that, besides being of general interest to the naturalist in this country, the subject might be made illustrative of a great natural law, and of the causes and process of distribution, and the increase in population of a species. I found also that while interesting to the naturalist and sportsman, it might also develop questions of economic value to Scottish or other landed proprietors, and be made to contain many local references interesting to individuals who own Capercaillie-haunted woods and forests.

With this view, therefore, I caused to be printed a series of queries touching the points I specially desired statistics and information upon, and by the beginning of 1878, I was engaged in sending out the circulars, receiving answers, and, as far as possible, arranging the results. The final results have far exceeded in interest my most sanguine expectations; though whether I have succeeded in conveying these results successfully and practically to my readers in the following pages it is for them to judge, not me. In the treatment of the subject I cannot lay any claim to originality, nor am I aware that any previously unknown facts are recorded. The treatment has been forced upon me by the large accumulation of data kindly put at my disposal by my many correspondents in this connection. The general remarks are the outcome of the statistics, and contain, I believe, little or nothing that is not known or believed by some of the landed proprietors and sportsmen in the area of the country inhabited by the species. I have not attempted to give my authority for every statement, where such a large amount of statistics came to hand. Errors no doubt must have crept in in such a compilation, for various reasons, but I believe that I have an authority for every statistic recorded.

I desire in this place to thank those who have so liberally assisted me. Had it not been for their interest in the subject, their never-wearying and hearty co-operation, and their courteous replies to my inquiries, it is needless to say this Essay could not have laid claim to the minuteness of detail which I trust it will be found to possess. Where all have assisted so liberally it might seem invidious to particularise;

but I cannot neglect to record my special thanks to a few who have apparently spared no trouble in assisting to gather materials, often from comparatively large areas, thus saving me an infinitude of time and labour, and a vast amount of personal investigation. I am also indebted to many others for assistance in the searches through old books in tracing the early history of the species in Scotland, and to several Gaelic scholars for assistance as regards the origin and etymology of the word "*Capercaillie.*"

To Sir Robert Menzies, Bart., I am obliged for some interesting notes of the earlier movements of the birds at Taymouth, and for other information. Colonel Drummond Hay of Seggieden; Sir Thomas Moncrieffe, Bart.; C. T. C. Grant, Esq. of Kilgraston; Dr. MacIntosh of Murthly; Mr. R. Paton of Perth; and Mr. R. Anderson, Dunkeld, have greatly assisted me in their various districts. James Haldane, Esq. of Cloanden, A. Burn-Murdoch, Esq., and J. Buchanan Hamilton, Esq. of Leny, have also collected materials for me between Perth and Callander. J. J. Dalgleish, Esq. of West Grange, has assisted me in the south of Perthshire; James Stirling, Esq. of Garden, and others, in the south-west; Mr. Dayton of Lochearnhead Hotel, and David Carnegie, Esq. of Stronvar, in the west and in Glen Dochart. My friend, W. Horn, Esq., has collected statistics from various parts of the Tay Valley, and also from other counties. Besides the above there are many others, representing over 150 estates in Perthshire alone, from whom I have received returned and filled circulars. To all I desire to express my best thanks.

In Forfarshire I am indebted to various correspondents. Especially I should mention W. Scott-Elliott, Esq. of Fotheringham and Tealing; W. Horn, Esq., for the Brechin and Stracathro district; and many others too numerous to mention, representing some 20-30 estates in the county.

In Fifeshire my thanks are specially due to J. Purvis, Esq. of Kinaldy, Charles Kinnear, Esq. of Kinloch, William Baillie Skene, Esq. of Pitlour, J. J. Dalgleish, Esq. of West Grange, J. Horne, Esq. of Thomanean, Robert Tullis, Esq., and others.

In Kinross-shire I am obliged to David Syme, Esq., Sheriff of Kinross, and to Messrs. Bethune, Henderson, and Burns Begg, for very full particulars; also to Harry Young, Esq. of Cleish.

In Stirlingshire, James Stirling, Esq. of Garden, T. G. Dundas, Esq. of Torwood, Sir James Gibson-Maitland, Bart. of Sauchie, T. Bolton, Esq. of Carbrook, and others, representing all the more important localities in the county, have assisted me.

In the south of Scotland I am indebted to several correspondents, amongst whom I may mention the Rev. James Porteous of Ballantrae.

In Arran I received a very full account of the restoration there from Mr. George Croll.

In Argyleshire, amongst others, I would mention the Rev. Alexander Stewart of Nether Lochaber, for much interesting matter.

In the north of Scotland my friend Thomas Mackenzie, Esq., Sheriff of Dornoch, has materially assisted me in Sutherlandshire; and the Rev. George Gordon of Birnie, and Captain Dunbar Brander of Pitgaveny, in Elgin and Banff.

To the Factors, Foresters, Gamekeepers, and many others who have sent me returns from single estates or localities, I desire to express my best thanks. It was upon these single returns that I depended in great measure for assistance in tracing out the *steps of advance;* and from the letters of my many obliging correspondents I often culled interesting additional facts.

I may mention that I have distributed more than 450 circulars to parties situated in all parts of the area at present occupied or visited by the species. By far the larger portion of these have been returned with the answers filled in, more or less fully, according to the amount of information my various correspondents had to impart.

Finally, Robert Warren, Esq. of Moyview, Sligo, Colonel Edward H. Cooper, of Markree Castle, Sligo, and A. G. More, Esq., Dublin, have rendered me valuable assistance, some of the results of which will be found in the Appendix, as they arrived somewhat too late for insertion in the text: and to my friend Mr. Robert Currie I am indebted for the careful and beautiful execution of his subjects, in illustrating the text, with the frontispiece, vignette, and other pieces.

PART I.

DERIVATION OF THE WORD "CAPERCAILLIE."

CHAPTER I.

DERIVATION.

THERE can scarcely be any doubt that the word *Capercaillie* or *Capercailzie*, with all the many variations in spelling,[1] comes direct from the Gaelic. The opinions of authorities, nevertheless, seem to differ in no small degree as to the correct Gaelic origin of the word, and much confusion exists as to the true meaning.[2]

This paper would perhaps scarcely be considered complete under its title if some notice of these different opinions were not taken, but the present writer having no knowledge whatever of the Gaelic language, must be satisfied with simply stating these views, without committing himself to any one of them.

The Rev. Dr. T. Maclauchlan, in a letter to Rev. G. MacArthur, 25th March 1876, holds the following views,

[1] See further on, p. 2.

[2] The old Scandinavian name of the Capercaillie, as I am informed by Dr. Meves of Stockholm—*fide* Dr. T. W. Lindblad, editor of the '*Swedish Hunting Journal*' (Nija Yagore Forbundets Tidskrift; Stockholm)—"is *thiàdur* (in English spelling, *chaidur*), often with the appendix *tupp*, denoting the cock, the he-bird. That, again, has a great many provincial and local variations, such as *tjådår*, *tedur*, *tjalur*, *tjuder*, *tjòddur*, in Dalecarlia—*tidder*, sometimes in Upland—and so on. In Vermland sometimes *tjúr-lian*, denoting the rich plumage or feather-garb of the cock. The Norwegian name is *tiur*. This word is supposed to be an onomatopoetic word, imitating as far as possible the peculiar 'playing' sound of the Capercaillie" (*in lit.*)

and his reputation as a Gaelic scholar and voluminous Gaelic author entitles these views to the highest consideration:—

About the second part of the word Dr. Maclauchlan considers there can be little room for doubt, and most Gaelic scholars appear to agree in this; but the first part of the word, he acknowledges, is more difficult. He says—"'*Cabhar*,' pronounced '*Cavar*,' means, according to our dictionaries, *a hawk* or *old bird*. It is not at all unlikely that it is the word spelled '*Caper*.' There is a similar word used in the name for a snipe, '*Gabhar-athar*,' thought by some to mean *the goat of the air*, from its bleating note. But," Dr. Maclauchlan continues, "it is a masculine noun, and '*gabhar*,' a goat, is feminine. I therefore lean to the idea that both in *Cabhar-athar* and *Cabhar-coille*—the one being *the bird of the air*, and the other *the bird of the woods*—the original term is *Cabhar*." Dr. Maclauchlan considers that "*Caber-coille*" is the orthography which comes nearest to the original. In a later letter to Professor Newton—who at that time was preparing an article on the Capercaillie for the *Encyclopædia Britannica*, and who has kindly put the above correspondence at my disposal—Dr. Maclauchlan states that the word *Cabhar* is not one in common use, and that "we are indebted for its meaning to our dictionaries, except in so far as it may enter into the formation of words like *Capercoille*. The Latin *senex*, so far as I apprehend, comes nearest to the meaning of '*old*' in cabhar, 'not *antiquus*.' There is a playful way of applying such words to the formation of names in Gaelic. For example:—*Bodach* is *an old man*, and *Bodach-ruadh, the red old man*, is the rock-cod. *Cailleach* is *an old woman*, and *Cailleach-aidhche, the old woman of the night*, is the owl. I think the Cabhar in this case is similarly applied."

Professor Newton (*Encyc. Brit.*, art. "Capercally") says:—"Cabhar, *an old man*, by metaphor *an old bird*, which is the acceptation of Dr. Maclauchlan's meaning = *the old bird of the wood*, the Capercaillie."[1]

[1] On the other hand, not a few Gaelic scholars consider that Capercaillie is derived from "*Capull, a horse*," see *capel, capell, caples*—Chaucer, line 170, 13-4—*vide* Bayley's '*Dictionarium Britannicum*' = *caballus*—"or, more correctly, *a mare*. *Capull* is a masculine noun, but at the present day is limited in its application to *a mare*, and *Coille, a wood*." This reading gives "*Horse of the woods*." In Argyleshire and Lochaber the bird is still known by the name *Capullcoille*. So also it is considered by several correspondents who are good Gaelic scholars. Amongst others, the Rev. Alexander Stewart of Nether Lochaber says:—"It is called 'Horse of the woods,' because of its size, strength, and beauty, as compared with other wood birds" (*in lit.*); and he further mentions that the name *Capullcoille* is found in Gaelic songs of the beginning of this century. The Rev. Lachlan Shaw, in his '*History of the Province of Moray*' (1775), also assigns this derivation: "properly, in Erse, *Capal coile, i.e.* The Wood Horse, being the chief fowl of the woods," (*op. cit.* p. 207). In Strathearn, in the south of Perthshire, where *native* Gaelic is now *almost* extinct, the name still lingers in this form. The first author of a Gaelic dictionary—M'Donald, an Argyle man—thus renders it, and all subsequent authors of Gaelic dictionaries do so likewise. Mr. D. Mackinnon, who has most kindly taken great trouble in this connection, looked up all the Gaelic dictionaries accessible, and informs me that all, without exception, give *Capull coille*. "None have *caper*, *cabar*, or *cabher*." . . . "The first Gaelic dictionary," Mr. Mackinnon informs me, "was written by M'Donald, an Argyle man, in 1741. Shaw, a native of Arran, prepared the next dictionary, and published it in 1780. Two small dictionaries were published in the latter part of the century by two Macfarlanes. In this century our two standard dictionaries—Armstrong's, a Saxon domiciled in Perth, and the Highland Society's, prepared by scholars from all parts of the country—were published in 1825 and 1828 respectively. There followed them M'Leod and Dewar's, two clergymen from different parts of the country; M'Alpine's, an Islay man; and M'Eachan's, a Roman Catholic priest, who spent his life, or the greater part of it, in Braemar. The only Irish dictionary I turned up has *Capullcoille*, quoted from Shaw. In the Scoto-Irish Dictionary, given in Llhuyd's '*Archæologia Britannica*,' the word does not appear."

Besides the above opinion, we have other derivations given. Jamieson, in his '*Dictionary of the Scottish Language*'—Supplement, 1825,—has as follows:—"Capercailye—yeane." A literary friend in the North of Scotland views Capercailye as compounded of Gael., *Cabar, a branch*, and *Caolach, a cock*. [Jamieson quotes the Scotch translator of Boece—Bellenden—here: "Gaelic, *Caolach*; C. B., *Kelliog*; Corn., *Kulliog*; Arm., *Kiliog*; Irish, *Kyleach*, a cock;" by which another element of confusion is introduced.] Cabar also means *an eminence*, or *the mountain*, which may have led writers astray in talking of the Capercaillie as specially "inhabiting mountains" (*v.* Burt, Ray, and others). Jenyns gets out of the difficulty by saying "mountainous forests." (I presume he used "forests" in the usual sense, and not in that of "a forest or chase.")

We are not yet done with combinations, as we have *Capullcaolach,* Horse-cock; and Yarrel, '*British Birds*' (1st edition), seems inclined to entertain this view, and finds parallels in "Horse-mackerell," "Horse-fly," (?) "Horse-leech." Or in German, "Auerhan;" Dutch, "Ouerhan;" or the Latin, "*urugallus*" (*urus,* a wild bull). Thus, again, Bull-finch, Bull-trout, etc., but some of these names can be traced to other sources than pre-eminence in size or strength; at all events the origins of such compounds are not by any means always clear enough to admit of generalisation.

There are other combinations, but the above-given—along with Dr. Maclauchlan's—appear to be the most important. In order to obtain the correct translations of Gaelic names, we must not, I believe, go to the Gaelic scholar *alone,* but first to the shepherd or crofter, whose family has for generations lived upon the same land, and whose father or grandfather was very likely the person who first applied the names, and which, being handed down from father to son, would preserve their purity of pronunciation, intonation, and significance, as well as, probably, a relation of the circumstances under which they were so named.

CHAPTER II.

ORTHOGRAPHY.

WE now come to consider the Orthography of the presently accepted word — *Capercaillie*, *Capercailzie*, or *Capercally* — with all the minor variations in the spelling of the terminal half—caillie. We may accept it as coming from the Gaelic *coille*.[1]

[1] Some of these variations are (*a*) *Capercailye;* first used by Bellenden in his translation of Hector Boetius in 1553. Bellenden has always been looked upon as one of the best—if not *the* very best—authorities on the subject of good old Scotch. He is followed, as late as 1808, by Dr. Jamieson in his *Scottish Dictionary*, and also in the *Supplement* in 1825, with *Capercalye* and *Capercalyeane.* (*b*) *Capercailles.*—This is used by John Graham Dalyell in his edition (1813) of '*The Chronicles of Scotland*,' compiled from the original manuscripts, going back as far as 1436—from which date the chroniclers continued the accounts of the Latin author—Hector Boece—and his subsequent translator; and Mr. Dalyell is careful to "preserve the old spellings" used in these manuscripts, which cannot be said to have been the case in *preceding* editions. In the second edition, however (1749), we find the spelling *Capercaillie.* This last is also used by King James VI. in 1617 (see '*Old Stat. Acct. of Scotland*,' xx. 473), and has been adopted by later writers, as Hewitson—'*Eggs of British Birds*' (1856), p. 277. (*c*) *Capercaile* (*v.* Foster, '*Synoptical Cat. of Brit. Birds*' (1817), p. 19. (*d*) *Capercaly*—Blaine, '*Encyc. Rural Sports*,' p. 82; and *Capercallie.* (*e*) *Capercali*—Lloyd's '*Game Birds of Sweden.*' (*f*) *Capercally* and *Caperkally*—Newton, '*Encyc. Brit.*,' ninth edition, quoting Pennant; and A. G. More, '*Ibis*' (1865). I cannot find Pennant's authority in the old law books for the use of this spelling, but old scribes sometimes spell a proper name two or three different ways in the same page. (*g*) *Capercalze* is first used by Bishop Leslie (Ed. of 1675, and also in the earlier edition of 1578), and was again used in Scots Acts, James VI., 1621, xxx., with the variation, *Caperkailzeis;* and this is followed by other

To begin at the beginning.—Some people assert that to spell it with a *z* is the best Scotch, but I am of opinion that, there being no *y* nor *z* in Gaelic, and the word being distinctly of Gaelic origin, it is best to adhere in form as closely as possible to that origin. But granting, in the meantime, the admission of *y* or *z*, then I think if the *z* be used it ought to be silent, as in many other Highland, or, I should say Scotch, names taken originally from the Gaelic, such as Menzies, Monzie, Colquhalzie (pronounced Cō-whȳ-lie), Rohalzion, Dalrulzion, Dalziell (also still spelt Dalyell), and others, mostly proper names.[1] *Because,* in old printing, a *z* was constantly used instead of a *y*, in proof of which, in the old Scots Acts of Parliament (see James VI., 1621, Act xxx.), the word "*years*" is spelt "*zeiris.*" In the self-same Act occur the words "*caperkailzeis*" and "*quailzies.*" This originated doubtless in a printer's error in reading the MS., or if not an error, then because *y* and *z* in old type were generally used as the same

writers, with other slight variations—*v.* Sibbald, '*Scot. Illust.*,' p. 16. Forster, (*op. cit.*) Blaine, (*op. cit.*) MacGillivray, *Capercailzie.* (*h*) Burt '*Letters from N. of Scotland,*' 1754, vol. ii. p. 173, uses *Cobber-kely,* pointing to a derivation from *Cabar*—as will be seen further on, or resulting from complications of local Gaelic dialects. (*i*) *Capercalcg* is used by Sir Robert Gordon in his '*History of the Earldom of Sutherland,*' (1630, published in 1813). (*k*) Sibbald, '*Scot. Illust.*,' (*Tables* 14, 18), Latinizes the word thus—*Capricalca,* following no doubt an earlier author, Robert Edward, ('*A Description of Angus,*' translated from the original Latin of Edward, minister of Murroes : Dundee, printed by T. Colvill, 1793), the word in the original being *Capricalcis* (ablative plural). For further notice of works quoted, see further on, p. 13. A very full list of the names it has received, and of the spellings used, will be found in the '*Penny Cyclopædia,*' in a very good article on the species, vol. vi., p. 260. Another variation—*Capercayllie*—is found in a '*Treatise on the Game Laws of Scotland,*' by A. Gregor, 1837, p. 9.

[1] There are other words which possess the *z* unpronounced, such as the law-terms *assoilzie, spuilzie, tailzie,* and others. Jamieson's *Dictionary,* however, does not give the z in *tailzie*—but '*taile, tailye,* an entail; *tailyie,* a piece of meat.' The fact of Jamieson omitting the z and substituting y, points to his knowledge that the z became inserted owing to the scarcity of the letter y in former and older founts of type. For the derivation of Monzie, see '*Old Stat. Acct. of Scotland,*' vol. xv. (1795), p. 241.

letter; or because, in default of *y* being sufficiently represented in the fount of type, the *z* was commonly used to replace it, or *vice versa.*

I may even go so far as to say that it was probably in this way that the *z* came to be introduced into all the above proper names, and into many others besides. There being no *z* in Gaelic, it has no right to appear in words of Gaelic origin: originating in error, it cannot be considered as good Scotch. Indeed, spelling the word with either a *y* or a *z* is perpetuating an error, arising from a misconception of a Gaelic word by a Lowlander. Even Bellenden—good authority as he is justly considered as an author of *Scotch* writing—cannot be followed in his spelling of this word, viz., *Capercailye.* In Gaelic the *ll* is liquid, like the French *lle;* and, as explained to me by a good Gaelic scholar—D. Mackinnon, Esq., of Edinburgh—when pronounced slowly, the sound might fairly be represented in Scotch by *lyie.* As the Rev. Mr. Macfadyen has pointed out to me (*in lit.*) this pronunciation can best be accomplished by dwelling on the second *l*, and having the tip of the tongue "not touching the roof or palate, close behind the teeth, but about an inch farther back." This gives the *y*-sound in the Gaelic; whence the insertion in MS. by a Lowlander of the *y;* whence, subsequently, the substitution of a *z* by the old printers. Many Highlanders at the present day pronounce the last half of the word as in the French word "*caille*"—a quail; and one individual, when saying it, added—"without the '*èh,*' which some give it." Mr. Macfadyen's name, as he himself shows me, is another good illustration of the insertion of the *y* and *z*. He says, "I spell it '*Macfadyen,*' but others '*Macfadzean;*' but in old registers it is without the *y* or *z*, thus—'*Macfadean,*' no doubt the original and correct spelling." Even the name Mackenzie in the old Gaelic name has no *z*. It was *MacConnich*—*nn,* like *ll,* being liquid and prolonged—Connich being the Gaelic name of Kenneth, and Mackenzie, as now used, being really the *son of Kenneth.*

But since the above was written, Professor Newton of Cambridge has called my attention to the fact that *y* and *z* were used, the one for the other, long before the days of printing; "and old English MSS. have," he goes on to say, "a mysterious letter ȝ or q, about the pronunciation of which some of the best old English scholars are in doubt; for in some words it is modernised into *gh*, if I remember right, frequently into *y* consonant, and less commonly into *z*."

Though this may appear at first to nullify my remarks on the interchange of *y* and *z*, still I think it cannot do away with them altogether, nor can it alter materially the fact that there being no *y* nor *z* in Gaelic, these letters should not occur in *Capercaillie*, unless, as Professor Newton suggests, as a terminal letter, thus—*Capercally* or *Caperkally* (plural, *ies*), for the English method.

A good illustration of the MS. use of the letter *z* is *cnizt*, knight. In the *Bannatyne MS.*, written in 1568—as I am informed by Mr. J. B. Murdoch, Glasgow—there are many *z* characters, which, on the authority of Mr. Thomas Dickson, Curator of the Register House, Edinburgh, ought to be rendered as *y*. In the same MS. y is used invariably for *th*. "Therefore," Mr. Murdoch adds, "I think the z had to be used for *y* where *y* was intended." According to some of the best authorities on old English MS., however, the use of *z* seems to be variously and differently applied from the use of it in Scotch MS.,—both as *y*, *gh*, or *z*, or *s*, according to its position in the words. In French the use is for *z* alone. In some words in old English MS. it stood also for *g*, (*vide* Morris and Skeat—"*Specimens of Early English.*" New and revised edition. Part II. Oxford, 1872).

Capercally or *Caperkally* may perhaps be looked upon as the correct English mode of spelling the word, if we take

Pennant as the first educated person who wrote it down. On this point Professor Newton writes (*in lit.*), "Pennant, who seems to have been the first British (as opposed to Scottish or Irish only) naturalist who mentions it as a bird of this country, says that it was called 'in the old law books *Caper-kally*.'" . . . I have hitherto in vain searched for this reading. "I suppose," continues Professor Newton, "he intended the second *a* to be sounded broad;" and if so, then, doubtless, the spelling will be the most correct English mode. We have seen, however, that Bellenden, as a good Scotch writer, spells it *Capercailye*. We have seen why the *y* should be dismissed. We find it dismissed by another excellent Scottish writer a little later, who compiles "from original manuscripts," and is an authority upon Scottish writing; and he writes "*Capercailles*," which is again changed by King James VI. into "*Capercaillies*," which I think should be accepted as the most correct Scotch way of spelling it, and approaching, moreover, nearest to the Gaelic pronunciation, as already explained.

PART II.

PRE-HISTORIC REMAINS,

AND ACCOUNT OF THE SPECIES IN SCOTLAND PRIOR TO ITS EXTINCTION.

CHAPTER III.

ANTIQUITY OF THE SPECIES, AS SHOWN IN BONE CAVES.

Of the antiquity of the Capercaillie we have evidence amongst the kitchen-middens of Denmark, where, as has been shown by Mr. Lubbock (*'Nat. Hist. Review,'* Oct. 1861, and Jan. 1862), and by Professor Newton (*'On the Zoology of Ancient Europe,'* 1862, p. 10), some bones of this species were identified. It is shown also that these kitchen-middens must be of great age, as the present old beech-groves of Denmark must have been preceded by an epoch of oak forest, "extending over several generations of trees," and that this oak-epoch was preceded by the era of pine, which alone could have supported the Capercaillie. The discovery of these Capercaillie remains created almost as much interest as that of the bones of the Great Auk (*Alca impennis*) in the same kitchen-middens.

Remains were also found in the caves of Aquitaine (*'Encyc. Brit.'*, 9th ed., vol. v. p. 53); (*'Reliquiæ Aquitanicæ'*). Professor Newton has quoted for me the passage in full:—"The Capercaillie is rare in the caves. I first recognised its remains in the station of Salève (near the shore of the Lake of Geneva), and at Verezzi (in Liguria). Afterwards I found this species at Bruniguel and Lacombe-Tayac; but it is represented there by only a small number of bones."

The fact is not to be passed by without remark that there

appears to be no trace whatever of remains of Capercaillies in Scottish kitchen-middens, nor amongst any pre-historic remains. Birds' bones are comparatively rare in these ash-pits, and this may be accounted for by their having been destroyed by dogs. Still, some birds' remains are found occasionally; and possibly, if attention be especially directed to them, some pre-historic trace of the Capercaillie may yet be found in Scotland or Ireland.

CHAPTER IV.

HISTORY OF THE SPECIES IN SCOTLAND PRIOR TO EXTINCTION; AND CAUSES OF EXTINCTION.

Of the occurrence of the Capercaillie in earlier historic times, and prior to the extinction of the species in Scotland, there is not much to relate which has not before been quoted by authors; but it may be desirable to place on record in a connected — and, as far as possible, a chronological — form, the more important notices.

1526. Hector Boetius first makes mention of the species ('*Hist. Scotorum. Scot. Regn. Descript.*', fol. xii. 47; and Bellenden's Translation, 1553).[1] To preserve the continuity of our account, I give it here in full. (It has been quoted before often. See MacGillivray; '*Hist. of Brit. Birds.*')

"Avium raptu viventium. Aquilæ sunt, Falcones, Accipitres, et id genus aliæ. Cæterum Aquatilium tam varius ingensque est numerus, ut pro miraculo notari haud ridiculum est. Sed medii inter eas quædam generis præter cætera reperiuntur aliis regionibus incognita. Unum magnitudine corvum paulo superans Auercalze, i. silvestris equi apelati, solius pinus arboris extremis flagellis victitantes." He then treats of Red Grouse and Black Game, with scarcely so much

[1] A later edition is "Scotorum Historiæ a prima gentis origine cum aliarum et rerum et gentium illustratione non vulgari, Libri xix. Accessit huic editione continuatio, per Joannem Ferrerium. Fol., Paris, 1574."

accuracy in his description, though MacGillivray's criticism is not quite correct either (*op. cit.* p. 143, lines 3, 4, and 5, from foot of page).

1528-29. In '*The Chronicles of Scotland*'[1] it is mentioned that King James V. "returned to Edinburgh," and the next summer (*i.e.* 1529) went to "Atholl to the huntis" (vol. ii. pp. 343-4). "The Earl of Atholl . . . maid great and gorgeous provisioun for him in all thingis pertaining to ane prince . . . . . . with fleshis, beiff and mutton, lamb, veill, and vennison, goose, gryse, capon, cunning, cran, swan, pairtrick, plever, duik, drake, brissel, cock,[2] and paunies, black-cock, and muir-foull, capercailles," etc. (*op. cit.* p. 345).

1578. Bishop Lesly ('*Desc. Reg. Scotiæ*'—which is dedicated to Pope Gregory XIII.—'*Kal. Januar.*' 1578) fixes a locality for the Capercaillie. The following is quoted for me by Professor Newton from the edition of 1675, *Roma*, p. 24:[3]

"In Rossia quoque Louquhabria (*i.e.* Lochaber) atque aliis montanis locis non desunt abietes, in quibus avis quædam rarissima Capercalze, id est sylvester equus vulgo dictu, frequens sedit corvo illa quidem minor, quæ palatum edenticum, sapore longe gratissimo delinit. Victitat ex solis abietis extremis flagellis: Alia avis est etiam in his regionibus numerosa, superiore minor hirsutis pedibus palpebris rubricantibus nostri gallum tesquorum dicunt."

[1] "*The Chronicles of Scotland*, by Robert Lindsay of Pitscottie, Edinburgh, 1814, by John Graham Dalyell." This is the best edition, as it was compiled direct from the old manuscripts, and retains the old spellings. In Dalyell's edition it is said to be a quotation from a later manuscript, and it is added "This passage bears strong evidence of interpolation." Mr. Thomas Dickson, Superintendent, Search Department, Register House, informs me that he "does not think it is mentioned"—*i.e.* the Capercaillie—"in the household books of James V., in which most of the viands then in use are mentioned." Extracts from this record are published by the Bannatyne Club, called "*Excerpta e libris Domicilii Jacobi V.*"

[2] "Brissel, cock" (*sic*), probably for "brissel-cock," or *coq de broussailes*, as suggested *in lit.* by Prof. Newton.

[3] The edition of 1578 has the above at p. 25.

1617. In a letter written by King James VI. to the Earl of Tullibardine, ancestor of the Duke of Athole, in 1617, "Capercaillies" are mentioned. As the passage is interesting, we give it in full:—"James, Right trustie and right well-beloved cosen and counsellor, We greet thee well. Albeit our knowledge of your dutiful affection to the good of our service and your countrie's credite doeth sufficientlie persuade us that you will earnestlie endeavour yourself to express the same be all means in your power; yet there being some things in that behalf requisite, which seem notwithstanding of so meane moment as in that regaird, both you and others might neglect the same, if our love and care of that our native kingdom made Us not the more to trie their nature and necessity, and accordingly to give order for preparation of every thing that may in any sort import the honour and credite thereof. Which consideration, and the known commoditie yee have to provide, *Capercallies* and termigantis, have moved us very earnestlie to request you to employ both your oune paines and the travelles of your friendis for provision of each kind of the saidis foules, to be now and then sent to us be way of present, be means of our deputy-tresaurer; and so as the first sent thereof may meet us on the 19th of April at Durham, and the rest as we shall happen to meet and rancounter them in other places on our way from thence to Berwick. The raritie of these foules will both make their estimation the more pretious, and confirm the good opinion conceaved of the good cheare to be had there. For which respectis, not doubting but that yee will so much the more earnestlie endeavour yourself to give us good satisfaction anent the premises, as yee will do us acceptable service. We bid you farewell.—At Whitehall the 14th Marche 1617." ('*Old Stat. Acct. of Scotland,*' xx. 473. See also under Parish of Dowally, farther on, p. 25.)

1618. In Taylor's '*Visit to the Brea of Marr*,' in 1618,[1] there occurs the following passage (p. 135):—

"Thus with extreme travell, ascending and descending, mounting and alighting, I came straight to this place where I would be, in the Brea of *Marr*, which is a large county . . .

"My good Lord of *Marr* having put me into that shape, I rode with him from his house, where I saw the ruines of an old castle, called the Castle of *Kindroghit*. . . . It was the last house I saw in those parts; for I was the space of twelve dayes after, before I saw either house, cornefield, or habitation for any creature but deere, wilde horses, wolves, and such like creatures. . . .

"Thus the first day wee traveld eight miles, where there were small cottages built on purpose to lodge in, which they call Lonquhards. I thanke my good Lord *Erskin*, hee commanded that I should alwayes be lodged in his lodging, the kitchen being alwayes on the side of a banke, many kettles and pots boyling, and many spits turning and winding, with great variety of cheere: as venison bak't, sodden, rost, and stu'de, beefe, mutton, goates, kid, hares, fresh salmon, pidgeon, hens, capons, chickens, partridge, moorecoots, heathcocks, *caperkellies*, and termagants. . . .

"All these, and more than these, we had continually in superfluous abundance, caught by faulcons, fowlers, . . . to victuale our campe, which conseisteth of fourteen or fifteen hundred men and horses. . . .

[1] "*All the Workes of John Taylor, the Water-Poet, Beeing Sixty and three in Number, Collected into One Volume by the Author: With Sundry new additions, corrected, revised, and newly imprinted*, 1630. *At London, printed by J. B. for James Boler, at the figure of the Marigold in Paul's Churchyard*, 1630. *Folio.*" At page 122 it is mentioned he left London

"The yeere of grace, accounted (as I weene)
One thousand, twice three hundred and eighteen,
And, to relate all things in order duly,
'Twas Tuesday last, the foureteenthe day of July."

I am indebted to Professor Newton for kindly transcribing this passage and title for me at length.

"Thus having spent certaine dayes in hunting in the Brea of Marr, wee went to the next county, called Bagenoch, belonging to the Earle of *Engie*. . . ."

1621. The species is mentioned in '*The Old Acts of the Scottish Parliament*,' notably in the reign of James VI., A.D. 1621 (Act. xxx.), where provision was made against the buying and selling of "wyld foulles," amongst which were included :—"termigantis, quailzeis, *caperkailzeis*, etc.," under a penalty of a hundred pounds.[1]

1630. Sir Robert Gordon ('*History of the Earldom of Sutherland, up to the year* 1630,'—not published, however, until 1818), mentions the species as at that time inhabiting the county. The passage containing the record is as follows :—

"All these forrests and schases are verie profitable for feiding of bestiall, and delectable for hunting. They are full of reid deir and roes, woulffs, foxes, wyld catts, brocks, skuyrrells, whitrets, weasels, otters, martrixes, hares, and fumarts. In these forrests, and in all this province, ther is great store of partridges, pluivers, *capercalegs*, blackwaks, murefowls, heth-hens, swanes, bewters, turtle-doves, herons, dowes, steares or stirlings, lairigigh or knag (which is a foull lyk vnto a paroket or parret, which maks place for her nest with her beck, in the oak trie), duke, draig, widgeon, teale, wild-gouse, ringouse, routs, whaips, shot-whaips, woodcok, larkes, sparrowes, snyps, blackburds or osills, meweis, thrushes, and all other kinds of wild foule and birds, which ar to be had in any pairt of this kingdome."

This passage has been quoted frequently before—see the '*New Stat. Acc. of the County*;' also '*Proc. Nat. Hist. Soc. of Glasgow*,' January 3, 1871; *Separate-Revised*, 1874, p. 69.

1651. In '*The Black Book of Taymouth*' (*Bannatyne Club Pubns.*, 1855), pp. 433-34, occurs the following passage, which

[1] Also in Act 1600, c. 23 (*fide* Irvine, '*Game Laws of Scotland*,' p. 5).

is partly quoted in Irvine's '*A Treatise on the Game Laws of Scotland,*' pp. 59-60 :—"To the Right Worshipfull, his much honoured freind the Laird of Glenorquhy, thes :—Much honoured Sir, Immediatlie after the receat of your letter on Saturday, I went and shew your Capercailzie to the king in his bedchamber, who accepted it weel as a raretie, for he had never seen any of them before." Signed "Jo. Dickson. Perth, the 3. of Februar. 1651."

1678. In a small pamphlet entitled :—'*A Description of Angus, translated from the original Latin of Robert Edward, Minister of Murroes—Dundee. Dundee: printed by T. Colvill —1793,*'[1] a passage occurs at p. 17, as follows :—"Angus is well stored with tame fowl and the larger kinds of birds, as hens of Brazil, peacocks, geese, and ducks. Pigeon-houses are frequent. The mountains and heaths abound with partridge, *grouse,** and plover, etc. etc." To this the translator adds the following *footnote* :—"* The word in the original, here rendered 'grouse'—or moor-fowl—is Capricalcis (Ablative plural). The translator could not find this word in any dictionary to which he had access." The translator also takes notice in the same *footnote*, that the last one that was seen was about thirty years ago in Strathspey, and adds :— "It is still an inhabitant of Sweden and Norway, and also of some parts of Wales." Here there is probably again some confusion; and as regards Wales, true grouse are no doubt intended.

1684. Sibbald, in '*Scotia Illustrata,*' 1684, includes the "Capercalze" in his treatment of the subject :—'*De ani-*

[1] In the Preface, or, as it is styled, "Advertisement," it is stated :— "The following description of Angus was originally written in Latin, by Robert Edward, minister of Murroes, and published in the year 1678, along with a pretty large map of the county, executed by the same hand." The only copy of the original was found—it goes on to say—about ten or twelve years ago, at the House of Panmure among some loose papers, etc. Dr. J. A. Smith, of Edinburgh, to whom I am indebted for the above extracts, informs me that the pamphlet is in the Signet Library. Edinburgh.

malibus Scotiæ,' p. 16; and in Tables 14 and 18, figures male and female, 'Capricalea.'

1754. Burt (*'Letters from the North of Scotland,'* 1754, vol. ii. p. 169), says:—"Of the eatable part of the feathered kind peculiar to the mountains, is:—First, the Cobber-kely, which is sometimes called a wild turkey, but not like it otherwise than in size. This is very seldom to be met with, being an inhabitant of very high and unfrequented hills, and is therefore esteemed a great rarity for the table." This record brings us down close upon its extinction in Scotland. Jamieson, in his later edition of Burt's *'Letters,'* adds a footnote:—"The Capercaillie, capulcoillie, avercailye, became extinct in Great Britain about this time, or shortly after," *i.e.* about the date of Burt's letter xxi. (1754?)—Edin. 1818, p. 71.

1775. In *'A History of the Province of Moray'* (1775), by Rev. Lachlan Shaw, 2d edition, 1827, p. 207, occurs a somewhat full notice of the Capercaillie, as follows:—"The harmless wild fowls are the swan, *Caperkylie* (called also Cock of the Wood); in Latin, Capricalea, as if he infested the goats; but properly, in Erse, *Capal-coil—i.e.*, the Wood Horse, being the chief fowl in the woods. He resembles, and is of the size, of a turkey-cock, of a dark grey, and red about the eyes; he lodges in bushy fir trees, and is very shy; but the hen, which is much less in size, lays her eggs in the heather, where they are destroyed by foxes and wild cats, and thereby the Caperkylie is become rare. His flesh is tender and delicious, though somewhat of a resinous fir taste."

1769. Pennant, in his *'Tour in Scotland,'* 1769, has an interesting passage regarding its occurrence prior to extinction in Inverness, which has been often quoted. He appears only to have seen one specimen, which "was killed in the woods of Mr. Chisholm, to the north [*in err.* for west.—J. A. H. B.] of Inverness."

Mr. Harting (*'Hand Book of British Birds,'* p. 38) says :— "One of the last native birds killed was shot at Chisholme Park, Inverness, and is believed to be in the Museum at Newcastle-upon-Tyne ;" but it would have been better had the grounds for this belief been stated. Later, Professor Newton (*'Encyc. Brit.,'* 9th ed., art. '*Capercally*') says :—"No British specimen known to exist in any museum"—*i.e.*, no specimen of the indigenous stock (p. 54). In reply to inquiries for further particulars, Professor Newton referred me to Fox's '*Synopsis of the Newcastle Museum,*' p. 78. On referring to the passage, I find that Fox was "unable to make out if the present specimen [*i.e.*, the male specimen in the museum.—J. A. H. B.] be really of British capture." Professor Newton, commenting on this specimen, writes to me:— "All that seems certain is, that the specimen at Newcastle was once Tunstall's, and that Tunstall, who was aware of the increasing rarity of the species in Scotland, does not say that he had a Scottish example; while he mentions one in his possession from Siberia, and also that he had had it from Denmark. This last, by the way, was most likely of Swedish or Norwegian origin, for the bird has been extinct in Denmark so long, that Steenstrup's discovery of its bones in a kitchen-midden was looked upon with almost as much interest as his finding the Garefowl's remains there" (*vide antea*, p. 13). Professor Newton further remarks (*in lit.*):— "Fox, I believe, is mistaken in considering the female Wood Grouse in the British Museum to be a British specimen from *Bullock's* collection (*l. c.*). It is entered in the B. M. catalogue as from *Montagu's* collection. Now, Montagu never mentions a Scottish specimen; and as in 1789—when the species was almost or quite extinct in Scotland—he was only beginning a provincial collection, it is most unlikely that he could have supplied himself with one. In Bullock's sale catalogue there is no evidence of his having a Scotch specimen, and he would know how much the value of his pair would have been enhanced had he been able to say they—or one of them—were Scotch. As it was, they sold for only £7, which, as prices went at that sale, was very moderate."

1776. Pennant, in his '*British Zoology*'[1] (4th ed., 4to, London, 1776, vol. i. p. 223), says:—"This species is found in no other part of *Great Britain* than the Highlands of Scotland north of Inverness, and is very rare even in these parts. It is there known (p. 224) by the names of Caper-calze, Auer-calze, and in the old law-books Caperkally." And in '*The Caledonian Zoology*,' in Lightfoot's '*Flora Scotica*' (1777), p. 22, Pennant further states that it is "found in the forests north of Loch Ness, but rarely. Once frequent in most of the Highland fir-woods."

1783. Latham ('*Synops.*' 11, p. 730) has as follows:—"The last bird of this kind found in Scotland was in the Chicholm's (*sic*) great forest of Strathglass; and I am well informed that the nest was placed in a Scotch pine." In the text, however, he states that the eggs are laid on the ground.[2]

'*The Old Stat. Acct. of Scotland*' has the parallel passage (vol. xx. 1798, p. 307):—"The Caper-coille, or Wild Turkey, was seen in Glenmoriston, and in the neighbouring district of Strathglass, about forty years ago; and it is not known that this bird has appeared since, or that it now exists in Britain."

[1] There appears to be no third edition of Pennant's '*British Zoology*,' for, as I am informed, the first came out in 1766 (folio); the second was begun in 1768 (8vo); and in 1776 there were two issues, one in 4to and the other in 8vo; and both bear the words "*Fourth Edition*" on their title-pages.

[2] That Latham's statement regarding the site of the nest may have been quite correct, is rendered possible by the following, which I extract from '*The Journal of Forestry*' for October 1878, p. 443:—

"*Capercailzie's Nest in a Pine Tree.*—In a wood adjoining Falkland House, on the Estate of Falkland, and at the northern base of the East Lomond Hill in Fifeshire, a Capercailzie has actually tenanted the old nest of a falcon in a pine tree, a considerable distance from the ground. The Capercailzie has laid ten eggs, and the process of incubation has advanced successfully in this aerial retreat."—"*Ornis*," *in* '*Land and Water*.'

1785. In Martyn's '*Dictionary of Natural History*,' 1785, it is said:—"The Cock of the Mountains is a native of various parts of Europe, and particularly of Ireland and Wales; but he is never seen in England except through mere accident." At another place he says:—"Capercalze—a provincial name for the Cock of the Wood." He describes the bird fairly well, "of the size of a peacock; appears to be of the pheasant kind!"

1802. Montagu ('*Dictionary of British Birds*,' 1802) did not appear to have been aware of its extinction in Scotland at the time he wrote, but quotes the above passage of Latham's regarding the nest found "placed in a Scotch pine." In his supplement, however (1813), he says:—"This bird, we believe, is now extinct in the British dominions."

The Capercaillie continued in Strathspey until the year 1745. The last seen in Scotland was in the woods of Strathglass, about thirty-two years ago (Rev. J. Grant in '*Old Stat. Acct. of Scotland*'—parish of Kirkmichael, Banffshire, vol. xii. p. 451). This would put the date of extinction about 1762. The above account is very generally accepted and quoted by later writers (see Fleming, '*Brit. Animals*,' p. 46; Gray, '*Birds of West of Scotland*,' p. 229, and others). Sir Wm. Jardine ('*Nat. Library: Orn.*,' vol. iv., 1834) puts the probable date later—between 1774 and 1784,—but we may accept the earlier record as being most probably correct. Professor Newton ('*Encyc. Brit.*,' 9th edition, article 'Birds,' p. 736, part xii.) places the extinction in Ireland at about 1760, and in Scotland "not much later," after comparing the accounts of Boece (1526), Bishop Lesly (1578), Pennant (1769), and all previous authorities which he had access to (*op. cit.*, art. '*Capercally*,' vol. v. p. 53).

'*The Old Statistical Account of Scotland*'—published between 1791 and 1799—contains other notices of the bird. Thus:—"The Caperkaily, or *king of the wood*, said to be a species of wild turkey, was formerly a native of this parish (Kiltarlity), and bred in the woods of Strathglass. One of these birds was killed about fifty or sixty years ago in the churchyard of Kiltarlity" (see '*Stat. Acct. of the Parish*'). The Capercaillie appears also to have inhabited the parish of Dowally, Perthshire (*op. cit.*) See also the letter written by James VI. in 1617 to the earl of Tullibardine, before mentioned (*v. p.* 17). Nearly all of the above records have already been compared and referred to by various authors; and Mr. Robert Gray ('*Birds of West of Scotland*') says:—"All records agreeing in the fact of the bird being extremely rare between the years 1745 and 1760, when it apparently became extinct." In '*The New Statistical Account*' of Perthshire (1841), it is mentioned as pre-existing in the parish of Fortingal (p. 543).

Yarrell ('*Brit. Birds*,' 1st edition) says—"There is even reason to believe that it (*i.e.*, the hybrid) formerly existed in Scotland, contemporary with the Capercaillie. Mr. G. T. Fox in his '*Synopsis of the Contents of the Newcastle Museum*,' published in 1827, quotes the Tunstall MS. at p. 78, in the following words:—"I know some old Scotch gentlemen, who say they remember, when young, there were in Scotland, both the Cock of the Wood and also the hybrid; and at p. 245, Mr. Fox has given a figure of this last-named bird, from a specimen in the Newcastle Museum . . ." But it is not said that this particular specimen was from Scotland. (Compare remarks under Harting—Newton, *antea*, p. 22.)

Eyton ('*Rarer British Birds*,' 1836, p. 30, footnote) mentions *T. urugallus* and T. *medius*, Meyer, as "formerly inhabitants of the British Isles, but are now extinct."

The hybrid is noticed as a native of Scotland by Brisson, under the name of '*Le coq de bruyére piquête*,' and, as we are informed by Fleming ('*Brit. An.*,' p. 46). "A Scottish gentleman told Dr. Tunstall, who informed Dr. Latham, that it existed in our woods."

Mr. A. G. More (*'Ibis,'* 1865, p. 426), while briefly alluding to the statements of Pennant and the *'Historia Scotorum,'* adds:—that the Rev. George Gordon told him that it also at one time inhabited the county of Elgin: also that Mr. Shearer "marked it as extinct in Caithness," implying thereby its former occurrence in that county.

As we have seen, the final extinction of the species may be considered to have taken place about the year 1760. There are, however, several other records of a considerably later date, which are at least worthy of notice. The latest I find on record are two given in Graves's *'British Ornithology,'* vol. i. By this account we find that "one was killed by a gentleman, of the name of Henderson, near Fort-William, about *six* years ago, and sent to Dundee . . . The other specimen was shot by Captain Stanton near Borrowstownness, *two* winters ago: they are both males. Some few are said to be yet remaining in the pine forests of Scotland, and also in the mountainous parts of Ireland." Now, the date of Graves's second edition was 1821, and Mr. J. H. Gurney junior informs me that he has a note to the effect that his first edition was published in 1817. The plate of the Capercaillie bears date of 1813 (*'Scot. Nat.,'* vol. i. p. 82). If we take the date of 1813, when he might have been expected to get his information, the Fort-William bird would be killed about 1807, and the Bo'ness bird in 1811.

With regard to the Fort-William bird, the Rev. A. Stewart ("Nether Lochaber," of the *Inverness Courier*) writes to me as follows:—

"The Mr. Henderson who killed the Capercaillie in the Camus-na-gaul woods, opposite to Fort-William, in or about the year 1807, was factor for MacLean of Ardgour, and tenant of the farm of Kiel, near Corran Ferry. At Ardgour House, about twenty years ago, the gardener was an old man of the name of Harry Kennedy. This Harry Kennedy was a man of superior intelligence, a naturalist and botanist of much and very exact knowledge in all that concerned these, his favourite studies. It was this old man Kennedy who first told me about the shooting of the Capercaillie by Mr. Henderson in 1807. Kennedy was then (1807) old enough to remember the circumstances perfectly. I am pretty sure that he said he saw and examined the bird after Henderson had brought it to Ardgour House, to show it as a curiosity to Colonel MacLean, the Laird."

We cannot look upon this male bird as a remnant of the original stock, but probably as a wandered bird from some later attempt at restoration, notwithstanding the mention of the Capercaillie in old Gaelic songs of the beginning of the century, which Rev. A. Stewart has quoted for me as follows, though there *does* appear to be traditionary evidence of its having lingered in that part of the Highlands, and, as Mr. Stewart says, that it was at least not an *unknown* bird in 1794. I give Mr. Stewart's communication in full :—

"In the year 1794, on the anniversary of the birthday of Campbell, Laird of Lochnell, near Oban, he gave a feast and ball to his tenantry and dependants. Present amongst others on the occasion was James Shaw, better known to Gaelic scholars as *Bard Loch-nan-Ealer* (the Lochnell bard). In the course of the evening, the bard having been called upon for a toast, repeated instead an extempore poem in laudation of his friend and patron the Laird of Lochnell, with which poem the Laird was so much pleased that he made Shaw a present of *Five Guineas* on the spot. The concluding line of one of the stanzas of this poem is this :—

'Bu tu *Capullcoille* na guibhsaich.'

and the literal translation of the stanza is as follows :—

'Thou art the eagle amongst birds,
Thou art the oak-tree not given to bending,

Thou art the salmon of silver-clear streams,
Thou art the Capercaillie of the fir-woods.'"

It only lived, I believe, *in the memory* of the bard. The tradition of its former existence was indirectly perpetuated in his poem.

There is good reason for believing that the Bo'ness record is so far genuine, and indeed that a bird, as recorded by Graves, was procured there; but from the situation of Bo'ness, upon the shore of the Firth of Forth, and its being a sea-port town, carrying on a trade in deals and timber with Norwegian ports, the probability exists that the male bird shot there may have escaped from, or been let loose by, sailors. Accordingly, this record, as well as the Fort-William one, must be received with caution as regards their real value, and all the more so that such a long interval exists between these and all previous records. Fleming, moreover, writing in 1828 ('*British Animals*,' p. 46), after mentioning the occurrence of the last birds in Strathglass (1860) and Strathspey (1845), says:—"Recent attempts have been made to recruit our forests [*i.e.*, with Capercaillies.—J. A. H. B.] from Norway, where the species is still common;" which attempts failing, and the cocks wandering, would be almost sufficient to account for these stray occurrences. At all events, in the absence of distinct data, it is safer to accept the date of 1760 as that of the extinction of the original stock in Scotland.

'*The Traveller's Guide, or a Topographical Description of Scotland*,'—Edinburgh, 1798—still speaks of the 'Caper-*ceilzie*' as existing at that date in Scotland (p. 4), but, as already seen, this is extremely unlikely.

Sir Robert Menzies informs me he "has always understood" that the last killed in his district—*i.e.*, along Loch Rannochside—was shot at Camagouran, by Gregor Macgregor, gamekeeper to the Laird of Struan, about the beginning of this, or the end of last, century; but in absence of further proof, I should be inclined to consider this an escaped bird from some one of the localities where the earlier attempts at reintroduction were made, as recorded by Fleming ('*Brit. Animals,*' 1828). Mr. Charles Buxton, editor of the '*Memoirs of Sir Thomas Fowell Buxton, Bart.,*' writing in 1852 (fifth edition, p. 332), states that the last bird "was shot about a hundred years ago in Perthshire," but this is probably an error (though *possibly* enough he may have had a record of a bird killed in Perthshire at that time[1]), and the record really most likely applies to the birds last seen in Strathspey or Strathglass.

There is no other evidence of a satisfactory nature that I can discover of the occurrence of the Capercaillie in Sutherland beyond Sir Robert Gordon's work already quoted; but Captain Houstoun, of Kintradwell, Brora, writes to me that one, Donald Sutherland (or Donald Mhor), used to mention the Capercaillie as having belonged to this county; but Donald Mhor, who died about twelve years ago, at the age of ninety—as the Rev. Dr. James M. Joass informs me—may have spoken from traditionary evidence, or of his father having seen it; or he may even have been quoting from Sir Robert Gordon's work.[2]

[1] Mr. Buxton may also have had in view the statements of its occurrence at Dowally in the '*Old Stat. Acct.*' (*loc. cit*), or that of its former occurrence in Fortingall Parish, given in the '*New Stat. Acct. of Perthshire*' (*loc. cit.*, p. 543), where it is stated that "we had *at one time* the Capercailzie (caper-coille), or great cock of the wood."

[2] The oldest pine trees in Sutherland are at Invercassley, on the opposite side of the river Cassley from Rosehall, and are now very few in number, as I am informed by my friend Mr. Thomas Mackenzie. "The Rosehall wood comes next, and dates from 1806. The Balblair wood, situated between Dornoch and Golspie, was planted about 1809—the same year that the small plantations of pine-woods about Kilcolmkill, in Strathbrora and Kintradwell, on the coast, were also put down in. These are the only standing woods in the county that have any pretensions to antiquity, the fir-woods on the Skibo estate being comparatively recent." Mr. Mackenzie comes to the conclusion that there is not a standing pine tree in Sutherland which is a hundred years old.

In Denovan's sale catalogue (1818) several passages occur, which have been kindly quoted for me by Professor Newton and Mr. J. H. Gurney jun., but I consider that these should be received with extra caution, or indeed be rejected altogether. As Professor Newton suggests, in those days there were tricks of the trade as well as now. I quote these passages for what they are worth:—

"*Lot* 651. Hybrid grouse. The rarest of the grouse tribe; was shot in Scotland; very fine; Edinburgh Museum.[1]

"*Lot* 652. Wood grouse; Highlands of Scotland. A handsome glazed case, including both sexes of these truly rare birds, in the finest possible condition."

To cap the above:—"*Lot* 832. Wolf, a noble animal, in large glazed case. The last wolf killed in Scotland by Sir E. Cameron."

1837. In a '*Treatise on the Game Laws of Scotland*,' by Alexr. Grigor, Edin. 1837, p. 9, "Capercayllies" are mentioned amongst pre-existing game birds. He did not, however, appear to be aware of their restoration.

IRELAND.

1357-87. Giraldus Cambrensis in his '*Topographia Hiberniae*' (lib. ii. p. 47), says:—"Pavones silvestres hic abundant." It remains uncertain if he alluded to the "Capercaillie" or not. We find again, however, that Ranulphus Higden, the monk of Chester, who died about 1360 (*v.* Harting on '*The Extinct British*

[1] In answer to inquiries as to whether any specimen at present exists in the Edinburgh Museum of a hybrid grouse or Capercaillie from Scotland that could have been referred to in the above lot, Mr. John Gibson assures me there is not. The oldest Capercaillie in the collection is a Norwegian one, purchased in 1814 by Dr. Jamieson for £4': 4s. Mr. Gibson considers that had there been a British example in the collection at that time, such a piece of extravagance would not have been perpetrated.

Wolf,' 'Pop. Sc. Review,' 1878, p. 397), in his *'Polychronicon,'* again makes mention of "pekokes" (pavonibus), which were probably Capercaillies (*v.* Translation, by John Trevisa, A.D. 1357-1387), quoted fully with remarks by Harting (*op. cit.*)

1676. Willughby (F.) (*'Ornithologiae Libri, etc.; recognovit Joannes Raius.* London : 1676'), mentions it as known in the country under the names "Cock of the Mountain or of the Wood," and "*Urugallus fœmina,*" as "The Hen of the Wood or Mountain," pp. 123, 124, and figs. on plate xxx.

1678. Ray, John (*'Ornithology,' etc.*, p. 173), says :—"This bird is found on high mountains beyond seas, and, as we are told, in *Ireland* (where they call it '*Cock of the Wood*'), but nowhere in England."

1684. O'Flaherty ("West, or H.-Iar Connaught") mentions the species under the name "Cocks of the Wood."

1772. J. Rutty (*'Nat. Hist. of the County of Dublin,'* 1772. vol. i. p. 302, *footnote*) says:—"Urugallus major, Gesneri: Tetras urugallus, Linnaei—'The Cock of the Wood,' British Zoology. One of these was seen in the county of Leitrim about the year 1710. But they have entirely disappeared of late, by reason of the destruction of our woods."

1760. Pennant also states that about 1760 a few were to be found about Thomastown in Tipperary, and Professor Newton (*op. cit.*) informs us that no later evidence is forthcoming, and adds :—"Thus it would seem that the species was exterminated at nearly the same time both in Ireland and Scotland."

Mr. Scoular, after quoting Giraldus and Act 11th Anne, adds that it "remained in the County of Cork till so late as 1750," on the authority of Mr. T. Whitla.[1] (*'Journ. Geol. Soc. Dublin,'* I. p. 227.)

[1] Other references to the species in Ireland will be found in the Irish Statutes, 11 Anne, ch. 7, which prohibits shooting of them for seven years.—Smith, '*History of Cork,*' 1749. The "Wild Turkey" of Act George III. 27, may or may not apply to Capercaillies, but the Rev. Dubourdieu's "Wild

An additional notice will be found in an able article '*On the Total and Partial Extermination of Animals*,' by James Fennel, in Rennie's '*The Field Naturalist*,' 1834, p. 194. This author says, that at that time the Wood Grouse or Cock of the Wood had been extinct in Ireland for nearly seventy years, and in Scotland for fifty years.

Mr. A. G. More of Dublin, who has been making inquiries in my behalf for some time past, regarding Capercaillies in Ireland, writes to me that he has not been able to obtain any additional information to the above in that country.

On the causes of the extinction of the species in Scotland I can say little. We can now, I believe, only speculate upon what changes of condition and what surrounding causes could have extirpated them. What appears to me to be the most likely factors were as follows:—The probable destruction of great forest tracts by fire,[1] the cutting down of the same by man as late as the days of Cromwell, and the wasting away of the forests from natural causes, by the conversion of dry forests into bogs and morasses, and, resulting from this, the decrease of, and changes in, the food of the species. Mr. Colquhoun ('*Ferae Naturae of the British Isles*,' p. 41-42) is of this opinion, but adds as a factor, the increased population. If we accept the above as the most probable causes, and come

Turkies" were undoubtedly "of the true breed" of Turkey from the American stock. For all the above references, see Thompson's '*Birds of Ireland*,' vol. ii. p. 31, from which I have quoted freely.

[1] Evidence of the destruction of great tracts of forest country are frequently to be met with in early history. Thus, to get rid of wolves, a large pine forest extending "from the western braes of Lochaber to the Black Water and mosses of Rannoch was burned to expel the wolves," and another "In the neighbourhood of Loch Sloi, a tract of woods, nearly twenty miles in extent, was consumed for the same purpose" (*v.* Notes to James Hay Allen's poem '*The Last Deer of Beann Doran*,' London, 1822). Sir Walter Scott also, in his Essay '*On Planting Waste Lands*' ('*Miscellaneous Prose Works*,' vol. xxi. p. 9), and other historians, afford evidence of a natural wasting away and decay of old forests, as well as their destruction by enemies, "thereby to remove a most important part of the national defence" (*op. cit.*, p. 10).

to examine into the details of the testimony, we find that it was not until the beginning of the 18th century that any large extent of young wood was planted,[1] nor until the end of the 18th century that arboriculture became general in Scotland. The latter would appear to have been too late to afford fresh sustenance to the indigenous Capercaillies, but it yielded an abundant supply by the date of the Restoration of the species in 1837-38, for the re-introduced birds. Rutty, as has been seen (*antea*, p. 31), accounts for their extirpation in Ireland "of late, by reason of the destruction of our woods." Smith ('*History of Cork*,' 1749) does so likewise.

[1] "It would seem that in Scotland, among the first who planted trees in large masses for profitable use was Thomas, Earl of Haddington. He began to plant extensively at his seat, Tynninghame, in East Lothian, in 1705. . . . . . A large amount of planting had been undertaken and performed betwixt the years 1777 and 1817. . . . In the latter part of the last century, and in the beginning of the present one, the chief planters in Scotland were the Duke of Athole, Lord Breadalbane, and Sir J. Grant of Strathspey. . . . . The plantations of any one of those just named far overbalanced those of any other private proprietor perhaps in Britain" (*v.* '*The Forester*,' by James Brown, 4th edition, p. 3, *et seq.* Edinburgh, 1871). Thus, coincident with the Restoration, it will be seen that there was abundant supply of young thriving wood at Taymouth. In '*The Beauties of Scotland*,' Edinburgh, 1805 (vol. i. p. 431), 1707, is given as the year in which planting was commenced at Tynninghame. A pretty full account of the operations is here given.

PART III.

RESTORATION.

CHAPTER V.

RESTORATION.

As has already been observed, Fleming, writing in 1828 (*'Brit. Animals'*), mentions that "recent attempts have been made to recruit our forests from Norway, where the species is still common."[1] It is probable that Fleming here alluded to the attempt made at Mar Lodge; as the previous year (1827), or the beginning of 1828, was the date of the first importation of these birds to Mar Lodge from Sweden (*vide 'Edinr. New Phil. Journal,'* July 1832). The account given (*op. cit.*) is very full and interesting, and I make no apology, therefore, for quoting it, as it will also serve to preserve the continuity of our account of the species in Scotland. Mr. Wilson writes as follows:—

"I was wading down the Dee one fine afternoon, a little below Mar Lodge, and with a lighter pannier than usual, when I heard the cry of a bird to which I was unaccustomed, and my bad success in that day's angling induced me the more readily to diverge from the 'pure element of waters' to ascertain what this might be. I made my way through the

[1] In 1829 some of the members of the Highland Society of Scotland proposed that the Society's attention should be directed towards the Restoration of the species, and mention is made of attempts having been made, "but," says the writer, "it is a difficult matter to alter the relations which naturally subsist between the wild animals of a country and the state of its population and surface."—'*Prize Essays and Transactions of the Highland Soc. of Scot.*,' New Series, vol. i. p. 5 (1829).

overhanging wood for a hundred yards, and soon after reaching the road, which runs parallel with the river on its right side, I observed a wooden palisade or enclosure on the sloping bank above me. On reaching it, I found it so closely boarded up that I had for a time some difficulty in descrying any inmates, but my eye soon fell upon a magnificent bird, which at first, from its bold and almost fierce expression of countenance, I took rather for some great bird of prey than for a Capercailzie. A few seconds, however, satisfied me that it was what I had never before seen, a fine living example of that noble bird. I now sought the company of Mr. Donald Mackenzie, Lord Fyfe's gamekeeper, the occupant of the neighbouring cottage. He unlocked the door of the fortress, and introduced me to a more familiar acquaintance with its feathered inhabitants. These I found to consist of two fine Capercailzie cocks and one hen; and the latter, I was delighted to perceive, accompanied by a thriving family of young birds, active and beautiful.

"The first importation of these Capercailzies arrived from Sweden about the end of the year 1827, or early in January 1828. It consisted of a cock and hen, but the hen unfortunately died after reaching Montrose Bay. As the male bird alone arrived at Braemar, the experiment was judiciously tried of putting a common barn-door fowl into his apartment during the spring and summer of 1828. The result was that she laid several eggs, which were placed under other hens, but from these eggs only a single bird was hatched, and when it was first observed, it was found lying dead. It was, however, an evident *mule* or hybrid, and showed such unequivocal marks of the Capercailzie character as could not be mistaken.

"The second importation likewise consisted of a cock and hen, and arrived safely in this country in January or February 1829. The female began to lay in the ensuing April, and laying, in general, an egg every alternate day, she eventually deposited about a couple of dozen. She showed, however, so strong a disposition to break and eat them, that she required to be narrowly watched at the time of laying, for the purpose of having them removed, for otherwise she would have destroyed the whole. In fact, she did succeed in breaking most of them, but eight were obtained uninjured. These were set under a common hen, but only one bird was hatched, and it died soon after. In the spring of 1830 the hen Capercailzie laid eight eggs. Of these she broke only one, and settling in a motherly manner on the other seven, she sat steadily for five weeks. On examining the eggs, however, they were all found to be addle.

"In the early part of 1831 three apartments were ingeniously formed adjoining one another. The hen was placed in the central chamber, between which and the enclosure on either side, each of which contained a male, there was an easy communication, so contrived, however, that the female could have access to both the males, whilst they, from their greater size, could neither approach each other nor disturb the female as long as she chose to remain in her own apartment. In May and June of that year she laid twelve eggs, seven of which were set under a common hen. Of these four were hatched in an apparently healthy state, one was addle, and the other two contained dead birds. Of those left with the Capercailzie hen she broke one, and sat upon the other four, of which two were hatched, and the other two were found to contain dead birds. Of the two hatched one soon died. Both the barn-door hen and the female Capercailzie sat twenty-nine days from the time the laying was completed till the young were hatched; and Mr. Cumming calls my attention to the fact that there were birds in all the eggs of this year's laying except one.

"My visit to Braemar took place about the first week of last August. I think all the five young were then alive, and although only a few weeks old, they were by that time larger than the largest moor-game. I had no opportunity of handling them, or of examining them very minutely, but the general view which I had of them at the distance of a few feet did not enable me to distinguish the difference between the young males and females. They seemed precisely the same at that time, both in size and plumage, although I doubt not the male markings must have soon shown themselves on the young cocks. The single surviving bird of those hatched by the mother died of an accident, after living in a very healthy state for several weeks. Two of those hatched by the common hen died of some disease, the nature of which is not known, after lingering for a considerable time. It follows that there are only two young birds remaining. These are both females, and when I last heard of them, some months ago, were in a thriving condition.

"The whole progeny were fed at first, and for some time, with young ants—that is, with those whitish grain-shaped bodies, which are the larvæ and chrysalids in their cocoons, of these industrious creatures, though commonly called ants' eggs. At that period they were also occasionally supplied with some tender grass, cut very short. As soon as they had acquired some strength they began to eat oats and pot barley, together with grass and the various kinds of moss. They are now fed like the three old birds, chiefly on grain and heather tops, with the young shoots and other tender portions of the Scotch fir. I am informed that the distinction between the sexes had become very obvious before the death of the young males. The plumage of the latter was much darker, their general dimensions were greater, their bills larger and more hooked. These characters became very apparent during November and December.

"The old males have never yet had access to the young birds, so that it has not been ascertained whether they entertain any natural regard for their offspring or would manifest any enmity towards them. From the continued wildness of the old birds, especially the males, it was found difficult to weigh them without incurring the risk of injuring their plumage. However, the male which arrived in 1829, and which then appeared to be a bird of the previous year, was lately weighed, and was found to be eleven pounds nine ounces avoirdupois. Judging from appearances, it is believed that the weight of the old hen would not much exceed one-half. There is, indeed, a striking disparity in the dimensions of the sexes in this species.

"The intention is, as soon as some healthy broods have been reared in confinement, to liberate a few in the old pine woods of Braemar, and thus eventually to stock with the finest of feathered game the noblest of Scottish forests."

Regarding the above-mentioned attempt to restore the species, it may not be out of place to make a few remarks. Since the above was first penned by Mr. Wilson, we know that the attempt ended in failure and complete annihilation of the birds, old and young. But subsequent attempts have—as will be seen farther on—resulted in their complete Restoration, and the experience thus obtained goes far to prove that the causes of failure in the Mar Lodge attempt were probably as follows:—

1st. The imported birds were too few in number, and the proportion of males and females not balanced as in nature, where the females of nearly all polygamous species exceed the males in population, and where, accordingly, one male is sufficient to serve a number of females.

2d. The confinement was probably against the complete health of the young birds, and, indeed, also of the old birds. Experience tells us that not one of the attempts made at Restoration, in which the birds have been too long confined has succeeded.

3d. The food would appear to have been not altogether suitable, if we accept Mr. Lloyd's directions for feeding Capercaillies in confinement ('*Game Birds and Wild Fowl of Sweden*'); and the general treatment in minor matters was probably deficient, from want of experience. The disease mentioned above by Mr. Wilson, of which one of the young birds died, and the deaths of the other young birds, were probably caused by some slight hitch in the general management of food and shelter. Even the wild birds, when young, are stated to be subject to 'gapes,' by more than one of my correspondents. Further, birds reared under domestic hens have never succeeded well; and we now know that the best way to treat the eggs is to place them out in the woods under wild grey hens, and to turn out the birds themselves to breed in a state of nature, after the woods have become partially stocked by the eggs hatched out under grey hens.[1]

It was a considerable time after the above attempt that any one earnestly took up the idea of re-introducing the species. "Years ago," writes Mr. Lloyd (*op. cit.*, p. 34), "I volunteered my services to more than one influential proprietor in Scotland. . . . For a long time no one would move in the matter, but at length, in the autumn of 1836, the late Sir Thomas Fowell Buxton, then recently returned from Taymouth Castle . . . took up the affair in good earnest." . . . "Influenced by the desire, in which I am sure you will concur,"—so writes Sir Thomas to Mr. Lloyd,—"to introduce these noble birds into Scotland, coupled with that of making Lord Breadalbane some return for his recent kindness to me,[2] I request you to procure for his lordship, at whatever cost, the requisite number." Sir Thomas also placed at Mr. Lloyd's disposal his head game-

[1] A further short account of the Capercaillies at Mar Lodge is given by Dick-Lauder in his '*Account of the Great Floods of August* 1829,' p. 358.

[2] *Vide* '*Memoirs of Sir Thomas Fowell Buxton, Bart.*,' fifth edition, edited by his son Charles Buxton, Esq., B.A. London, 1852. Chap. xxiv p. 332.

keeper, "Larry" Banvill; "which," as Mr. Lloyd remarks, "was no slight sacrifice for a Norfolk game-preserver."

Capercaillies were reintroduced at Taymouth in the autumn of 1837 and spring of 1838. In all, according to some accounts, forty-eight birds were obtained in Sweden, through the instrumentality of Sir Thomas Fowell Buxton, of Northrepps Hall and Cromer Hall, in Norfolk, and the energy of Mr. Lloyd, the well-known Swedish sportsman and naturalist, materially assisted by Mr. Lawrance Banvill, Sir Thomas Fowell Buxton's head gamekeeper in Norfolk, who went twice over to Sweden and brought over the birds to Taymouth, and by Mr. James Guthrie, head gamekeeper at Taymouth, who carefully tended the old, and successfully reared the young, birds.

According to a letter from Mr. Edward Buxton, published in Blaine's '*Encyc. of Rural Sports*,' (p. 814), a previous attempt to rear the birds had been made in Norfolk by Sir Thomas. A hen bred there, but all the young ones died. Several hens and a cock had been kept at Cromer Hall, where Sir Fowell was then living. This must have been about the year 1823, a few years previous to the publication of Mr. Lloyd's first large work—'*The Field Sports of the North of Europe*' (*q. v.*, vol. i. p. 264).[1] Mr. J. H. Gurney informs me that there is a hen bird stuffed in the Norfolk Museum, which belonged to Sir Fowell Buxton, and which was doubtless one of these birds, or one of those brought over later by Larry.

Mr. Henry Stevenson of Norwich gives me a further note from his journals, taken down by him from Larry's own oral communication, in the year 1857, to the effect that "a pair were turned off at Sir Fowell's place at Beeston, but soon died—he believes choked in some way. . . . Beeston, Cromer, and Northrepps, are all adjoining parishes, or within a mile or two of each other," so that it is not perhaps of great im-

[1] *Vide* also Loudon's '*Mag. of Nat. Hist.*,' vol. iii. p. 157.

portance to ascertain the fact exactly at which of the three places they were turned off. We may, however, accept "turned off at Beeston" as most likely correct.

Mr. Gurney also tells me that a journal left by Mr. Lawrance Banvill was destroyed by his son. It doubtless contained full notes upon Capercaillies, and it is to be regretted that it was not preserved.

In Blaine's '*Encyc. Rural Sports*' (1838), however, will be found a very full account of "Larry's" journey, in his own words, which is well worth perusal, but is too long for quotation in this place. A short abstract will, however, preserve the continuity of our account.

By "Larry's" own journal it would appear that he left home on the 1st April 1837, and made the Swedish coast on the 17th April, and on the 20th reached Mr. Lloyd's house at Venersborg. There he remained, collecting and attending to the birds, which were kept in large coops, five feet by three and a half feet, or thereby. Besides Capercaillies a few blackcock were procured, as well as a few eggs of both species. These birds were often brought in from a considerable distance, the natives having been instructed by Mr. Lloyd how to capture the birds without hurting them. Larry then details the troubles he encountered in finally preparing the birds and their coops for the voyage down the river to Gottenburg; but finally, on the 8th June, the birds were safely got on board, and Larry took—what he then believed, was—the last look of the Swedish coast on Saturday the 10th June.

On the 19th June Larry and the birds (in all thirty-six birds, being 7 black game and 29 Capercaillies, of which latter there were 13 cocks and 16 hens) landed at Hull. On the 21st he sailed for Dundee, where he arrived with all the birds well on the 23d. He left Dundee on the 24th, and appears to have reached Taymouth the same evening—a long journey, part of which was effected with a horse and cart. The birds were all well on the 25th, having reached Taymouth Castle in safety. Larry remained a short time at Taymouth, and got home again to Norfolk by the 20th July. Mr. E. Buxton, in his further account of the birds, sent to the proprietors of '*The Encyclopædia of Rural Sports,*' (continued, p. 816), gives an account of the successful hatching out of two broods in the woods, and the arrival of 16 more hens at Taymouth in the summer of 1838. Larry had again been over in Sweden assisting, as before, in their safe transmittal. In September 1839 Mr. James Guthrie, the head gamekeeper at Taymouth, wrote:—"With regard to the Capercaillies, I think we shall have between 60 and 70 young." Mr. E. Buxton also mentions the fact that already two strayed birds had been shot in the north of Scotland, and one hen had been offered for sale in a poulterer's shop in Princes Street, Edinburgh. I would recommend, however, to those interested in further particulars, to read the excellent account of the restoration in Blaine's work, of which the above is an extract.

Mr. Lloyd informs us that there were 20 birds in the second lot sent off from Sweden. The balance, I believe, were sent to Cromer Hall, but I have been unable exactly to trace what became of them, unless, as Mr. D. Brown, lately naturalist in Perth, informs me, they were taken not to Cromer, but to Northrepps, where young were reared; but all died, owing, Mr. Brown writes, to the want of proper food.

It would appear at all events, that, in all, 13 cocks and 29 hens arrived in safety at Taymouth in 1837 and 1838. Some accounts give 48 birds; and Knox, in his '*Game Birds and Wild Fowl,*' p. 221, says, "54 adult Capercaillies in 1838-39" (*fide* '*Zool.*' ix. p. 3018-9).

In September 1838 Sir Fowell saw eighteen of the birds at Taymouth in good health, and ten more had been turned down on the estate. The birds were carried out at night in large baskets, and placed about amongst the woods, principally around the castle; the lids were lifted, and the birds found their way out of their own accord. The actual rearing by hand was not so successful, but in 1841 favourable reports were received of the successful hatching of eggs under grey hens, principally in the woods of Drummond Hill. They soon became firmly established, and about the year 1862 or 1863 the Marquis of Breadalbane estimated their numbers on the estates at over 1000 birds, while the head gamekeeper, James Guthrie, who tended the birds with the greatest possible care, estimated them to reach over 2000. It is greatly owing to the intelligent care bestowed upon them by Mr. Lawrance Banvill and Mr. James Guthrie that this perfect success had been attained. To Sir Thomas Fowell Buxton and to Mr. Lloyd thanks are in the first instance due, but Messrs. Banvill and Guthrie, the Irish and Scotch keepers, must not be forgotten for the active practical part they took in the matter.[1]

About the same time that they were introduced at Taymouth five hens and a cock were brought to Dunkeld House for the then Duchess Dowager of Athole, but shortly afterwards the cock bird killed himself, and the hens were sent up to Taymouth.[2] In 1843 a successful introduction was effected in Arran from birds sent from Taymouth; supplemented

[1] Mr. Guthrie's "only delight or pleasure was to look after the 'beasties and birdies, puir things,' for which he had a great affection; but woe betide the vermin. The affection and humanity was of a different kind for the 'dirty vermin.' I have seen him hang a wounded crow to die over a caged hen with a brood of young pheasants, caressing the one, while to the other he was absolutely cruel; justifying the latter acts by the assertion that the one was 'vermin,' the others 'puir bonnie creaturs.' He came from Arbroath, when quite young, to Taymouth, and I believe the shooting interest in Perthshire owes his memory and zeal a debt, for he '*haated*' hawks, foxes, eagles, most sincerely, and spent three nights a week for years on the hill." I am indebted for the above 'memory' of an old friend, to one who knew him intimately, and who accompanied him in many of his long tramps o'er heath and hill and tangled knowe.

[2] Another account says, "three birds in 1838." Yarrell—'*Brit. Birds.*' First Edition.

by a fresh relay of birds direct from Sweden (see details under Arran) in 1846. Concerning an alleged independent restoration of the species at Murthly by Sir William Stewart, Bart., about 1844, I have good reason for stating that there is no actual foundation for the report, the birds arriving there as early as 1842, and coming of their own accord from the Athole woods. Mr. Malcolm Dunn, who has given me so much assistance under another heading (*infra*, p. 131), informs me that he is not responsible for the statement to the effect that they were there re-introduced, as stated in the '*Zoologist*,' 1875, p. 4338.

Mr. John Colquhoun informs me that he inspected the "colony of nests" at Taymouth. "The nests were close together, like rooks', and were thus more easily protected." In the fine autumn mornings he has often watched them feeding or disporting in full view of the windows of the house; and in spring, and again in autumn, this can be seen still in the quiet secluded parts of the forest, or even in more public parts of the estates.

For further details of the rearing of the birds see Yarrell's '*History of Brit. Birds*'—first edition—and Knox's '*Game Birds and Wild Fowl*,' p. 221.

In 1845 David Carnegie, Esq., of Stronvar, brought over two cocks and four hens, and kept them in confinement for some years, hoping to get eggs to place in grey hens' nests. The hens died in two or three years, and the cocks were let loose. The latter were found not long after, both dead. It was believed they had fought and killed each other. They had been enemies when in confinement, only separated by an open wooden partition (D. Carnegie, Esq. *in lit.*)

At Taymouth, and all along the Tay valley, as far as Dunkeld, Capercaillies, after becoming fairly established, increased in numbers rapidly for a number of years. The whole district was in every way admirably adapted to their habits, the Duke of

Athole and Lord Breadalbane having, as we have seen (*antea*, p. 33), planted considerable areas of their estates, in the latter part of the last century and in the beginning of the present one, with larch, Scots fir, and spruce, thus forming for the restored birds, the perfection of cover and food.[1]

Of late much of this wood has been cut down, and sheep-grazing has taken the place of forest growth. Consequently the birds find their domain restricted, and are more subject to disturbance. Mr. Anderson says :—" Within the last six or seven years they have been decidedly on the decrease, although still very abundant." All my correspondents agree in this, and one goes so far as to say that he " believes there are not half the birds upon the Athole estates that there were twenty years ago." I have statistics showing a similar decrease in other localities. At present the birds are not much shot nor disturbed on the Breadalbane and Athole estates.

Very shortly after their introduction at Taymouth they appeared at Craig-y-barns, near Dunkeld—viz. in 1840—and a female was shot in 1843. They were then preserved up to 1848, but had become firmly established there long prior to that date.

In 1842 Prince Albert shot one at Taymouth as part of the game obtained when there with Her Majesty.[2] Mr. Charles Buxton, in his '*Memoirs of Sir Thomas Fowell Buxton, Bart.*' (p. 333, *footnote*), claims to have shot, along with his brother, " the first of these birds that had been killed in Scotland for a hundred years." As has been related by Blaine, however, two were said to have been shot in the north of

[1] To Mr. Roderick Anderson of Dunkeld, amongst many others, I am indebted for a very succinct and admirable account of the reintroduction of the species; to Mr. Dayton of Loch Earn Head Hotel also, for other notes and hints connected with the subject, and others having had personal acquaintance with Mr. James Guthrie and Mr. Banvill, and who well remember the first appearance of the birds at Taymouth.

[2] *Vide* '*Her Majesty's Journal.*'

Scotland in 1839; wandered birds from Taymouth, probably, and one was offered for sale in Edinburgh.

In 1843, as I am informed by Sir Robert Menzies, "a large number of the birds," to use his own words, "came across the river to the Rock of Dull, on my property, where they bred very well, and were taken good care of for a number of years, as I did not allow them to be shot. The Rock of Dull is a large wood of old Scotch fir. It is well exposed, facing the south, and with plenty of water. The original plantation is about 300 acres, and there is as much again more recently planted close to it. With the exception of Drummond Hill, the woods at Taymouth face the north, and I presume that this was the reason why the birds, when turned out, came to my side of the valley, which was right opposite. At first they increased rapidly, having bred well the very first season; but I never got them to increase beyond what they were as to numbers in the second or third year. They are very wandering birds, and very soon I heard of their being at Tullimet, Dunkeld, and Blair Athole; and thence they shortly after moved on to the plantations in Strathord, Strathearn, and the upper woods in the Carse of Gowrie. It is noteworthy that this migration was always to the lower plantations, and I have made several unsuccessful attempts to turn them out again in Rannoch, which is higher wood. There, there is a quantity of the indigenous Scotch fir, the remains of the former Caledonian forest; and where I expected they would have done well and been well satisfied with their quarters, they remained a year or two and then disappeared, and, I think, migrated southward, preferring other places to Rannoch. I have also sent eggs on several occasions across the Grampians to Inverness-shire and to Beaufort Castle, but I think they have not succeeded as yet in introducing them there. I preserve the Capercaillie, as I do not think they do harm either to the grouse or black game; in fact, they frequent large well-grown Scotch fir plantations, where these are not usually found."

In Arran they were separately introduced—partly from birds direct from Sweden, and partly by birds from Taymouth. Notwithstanding the mixture of fresh blood thus acquired, and the corresponding rapid increase of the species there, Arran, from its isolated position, cannot be held as an appreciable *centre* for any large extent of country. (For full particulars of this introduction, *vide* under "Arran," p. 103.)

I am obliged to Mr. Muirhead, of Edinburgh, for the following statistics, which may perhaps be considered to represent the steady increase of the species since that time. He informs me that he received for sale in 1855 not more than 10 brace of Scottish birds; in 1865, from 15 to 20 brace; in 1875, from 20 to 25 brace during the season; in 1876, fully 30 brace; but in 1877-78 scarcely any. "This wet season," he adds, "partridges, grouse, and pheasants have turned out a very small crop, in fact, the shortest supply on record." Many Scottish sportsmen can bear out this latter remark of Mr. Muirhead's. In 1852 Mr. John Plant received a hybrid from Mr. Muirhead, shot near Breadalbane ('*Zool.*,' 1852, p. 3710).

In England also attempts have been made to introduce the species. Lord Ravensworth gives an account of his endeavours to establish them at his seat at Eslington in Northumberland, between 1872 and 1877, in the '*Nat. Hist. Trans. of Northumberland and Durham*' for 1877. At the latter date there were three males and two females in good health. In the south of Yorkshire, around Hebden Bridge, in the summer of 1877, six strong chicks were hatched out under the care of a Scottish gamekeeper, from eggs procured from the Highlands (*vide* '*Newcastle Chronicle*,' January 1877).[1]

[1] Mr. David Fyffe, gamekeeper at Lindertis, Forfar, sent seven young birds to the Duke of Newcastle. One cock grew to the size of a moderately large turkey; but through time they all died.

At an earlier date (1842) an attempt was made at Taplow Court, Buckingham (Thompson, '*Birds of Ireland*,' ii. p. 33); and several birds were sent to the Earl of Derby at Knowsley in the same year (*op. cit.*) The Zoological Society of London had a pair, but they did not long survive ('*Penny Cyclopædia*,' vol. vi. p. 260). By the following extract from the '*Auction Catalogue* (corrected) *of the Menagerie and Aviary at Knowsley*,' August 1851, p. 37,—copy in Library Zool. Soc., London,—it would appear that there were four birds at Knowsley at the time of the sale, October 1851 :—

Lot 517. (Divided into two lots.)	2 Capercaillie, male and female.			
	2	ditto.	ditto.	N. Europe.

The birds at Eslington all died off. The remaining male became too mischievous to be kept with safety, and had to be destroyed before the late lord's death.

Mr. Webb did not—as stated in '*The Field*' of January 2, 1875—turn out any Capercaillies at Newstead Abbey, Nottingham, and the hybrid reported (*op. cit.*) was a cross between a blackcock and a pheasant, *not* between a Capercaillie and black game ('*Field*,' January 23, 1875).

In Ireland, Lord Bantry attempted to introduce them at Glengariff. Three brace of birds were sent over by Mr. Lloyd, as related in Thompson's '*Birds of Ireland*,' vol. ii. p. 31. (*Vide Appendix.*)

PART IV.

INCREASE AND EXTENSION OF RANGE.

(With a Map.)

COPY OF CIRCULAR.

In re Capercaillie—Queries.	*In re Capercaillie—Answers.*
Name of Estate	Name of Estate.
County	County.
1st. Do they occur on above estate? .	1st.
2d. If so, are they resident or occasional?	2d.
3d. Were they introduced, or did they arrive of their own accord? .	3d.
4th. If introduced, when?—give year or (failing that), approximate date	4th.
5th. If introduced, where from and how—by eggs or by birds? . .	5th.
6th. If arrived of own accord, when? .	6th.
7th. And where *probably* from? . .	7th.
8th. When did they become fairly established on above estate? . .	8th.
9th. Are they "common," "abundant," or "rare"?	9th.
10th. How many have you seen in a day, or seen shot in a day; and when was it you saw this? . . .	10th.
11th. Give approximate extent of woods suitable for their habits—pine or fir—upon the above estate; in acres or square miles . . .	11th.
12th. Are they preserved?	12th.
13th. Are they increasing?	13th.
14th. Are Black Game decreasing since their arrival, or otherwise? .	14th.
15th. The names of adjoining estates where they occur, with any particulars you can furnish—in the above order	15th.
	Sign Name and Address.

N.B.—If you have any additional information, please put it on separate sheets of paper, and send by letter post.—J. A. H. B.

CHAPTER VI.

INCREASE AND EXTENSION OF RANGE.

In the following sketch of the increase and extension of range of the Capercaillie in Scotland under each county, I have given the distances in direct lines from Taymouth of each locality or group of localities mentioned. The actual lines of advance, however, must not always be understood to have followed these direct radii, because, as shall be further explained, the natural courses of the *waves* of *distribution*—if I may so term their acts of progress—follow, for the most part, the trend of the valleys up or down stream, or the wooded slopes of the hills, *seldom reaching across bare hills of any altitude.* The accompanying Map will, it is believed, greatly assist in indicating the lines of advance, and should be consulted whilst reading the text. Owing to the small scale of the map the names of many localities are necessarily omitted, but most of these places will be found upon Black's maps of Scotland, and, I believe, almost all upon the Ordnance Survey sheets, so far as published.

Explanation of the Map.—The map is on the scale of one inch to ten miles.

The map shows that portion of Scotland which may be said to contain the area at present occupied by the Capercaillie, *excepting* the Island of Arran—which may be considered as the seat of a separate introduction—and certain other outlying localities in the North, South, and West of Scotland, which I have treated of as fully as my materials admit of in the text, but which I have not thought it necessary to include in the map.

This area comprises Perthshire—the head-quarters of the species—Forfar, Fife, Kinross, Clackmannan, Stirling, and Dumbarton; and also the neighbouring portions of Argyle, Inverness, Aberdeen, and Kincardine, in the west and north; and the Lothians and south shore of the Firth of Forth in the south.

Those portions coloured an *uniform shade* of red show the localities where I have ascertained that the species is permanently established, or where the birds are at present *resident*. Those other portions, *dotted* red, show the localities where the species occurs *occasionally*, breeds *sporadically*, or only visits *at certain seasons.*

Taymouth, the CENTRE of RESTORATION, is marked thus, ⊙. The concentric rings show the distances from the CENTRE, on the scale of 1 inch to 10 miles, the scale of the map.

The dates of arrival of the species at the various localities are printed in black over the coloured portions. Where a capital E precedes the date, *Establishment* is recorded.[1]

Other localities where successful introductions have been effected—*i.e.* where Establishment has resulted—are also marked so, ⊙; but in the text Taymouth is always spoken of as THE CENTRE, whence all distances are measured.

Localities where unsuccessful introductions have been made are indicated by a cross, thus ×.

[1] While many of the dates given are exact, and have been in the first instance carefully recorded, many others must be accepted only as approximate. I found it impossible in many cases to obtain exact statistics of *Arrival* or *Establishment*, and many of the returns I have received contained no dates whatever. Such, I think, can hardly be looked for in all cases, but I believe by far the larger number of the dates and statistics given can be relied upon for all practical purposes.

MAP TO ILLUSTRATE
Extension of range of the
CAPERCAILLIE
IN SCOTLAND
Since its Restoration at Taymouth
1837 – 1838.
Scale of Miles
0
10
KINCARDINE
FORFAR
PERTH
STIRLING
FIFE
KINROSS
DUMBARTON
Forest of Athole
Montrose
Brechin
Arbroath
Carnoustie
Dundee
Firth of Tay
St Andrews
St Andrews Bay
Fife Ness
Crail
Isle of May
FIRTH OF FORTH
North Berwick
Kirkcaldy
Dunfermline
Burntisland
Leith
Falkirk
Glasgow
Aberfeldy
Dunkeld
Blairgowrie
Alyth
Killin
Callander
Doune
Lomond
Rannoch Moor
Shichallion
Ben Lawers
Ben Vorlich
Glen Almond
Glen Garry
L. Tunnel
Tyndrum
Dalmally
Lochearnhead
Kirriemuir
Glen Clova
Clova
Edzell
Forfar
Lunan B.
Red Hd
Ben Alder
Cruach
Crianlarich
Ben Ledi
Ben Lomond
Killearn
Buckhaven
Leven
Dysart
Broughty Ferry
Newport
Leuchars Jun.

CHAPTER VII.

PERTHSHIRE.

PERTHSHIRE continues to be the head-quarters of the species, and is not likely to lose its position in this respect for a long time to come, although Forfarshire is pressing hard to emulate it.

All the minute statistics I have been able to gather I have thought worthy of record, because there is no saying what changes of circumstance may affect the future census of the Capercaillie population, or what difficulties may arise at a later date, if the statistics of its earlier history since its restoration come to be desired.

I will now proceed to arrange, as best I can, the statistics, starting from Taymouth as THE CENTRE, and following the birds as they advance outward in steady "*waves of distribution.*"

1837 and 1838. Date of Restoration at Taymouth, marked ⊙ on map.

1840. Reached the neighbourhood of Dunkeld, one having been seen at Craig-y-barns (15 miles from ⊙), and a female at Langridge wood, near Scone (23 miles from ⊙). Now—1877-78—resident, and again increasing, though a decrease took place on the Athole estates for some years—preserved. This is part of a district on the Athole estates, stated roundly as containing 10,000 acres of suitable woods.

1841. The Capercaillie at this time had become fully

established at Taymouth. At present (1878) said not to be so abundant as twenty years ago. Mark in this connection the migratory movement related by Sir Robert Menzies (*antea*, p. 49).

1842. Arrived at Murthly, down the Tay valley (23 miles from ⊙, or 30 miles following the river). Were much more numerous here 15-20 years ago than they are at present, but are again on the increase now. They have 3000 acres of wood to thrive in.

1843.[1] One, a female, shot at Craig-y-barns (*ut sup.*)

1844. Became established at Logierait (14 miles from ⊙).

1845. Became established at Murthly (*ut sup.*) An attempt was made to introduce them at Stronvar in Balquihidder. Mr. David Carnegie brought over two cocks and four hens, and kept them for some years in confinement. (See under account of Restoration.) Stronvar is 24 miles in a direct line from ⊙, or thereabouts. A male bird was caught alive at Blair Castle (13 miles ⊙), and was sent back to Taymouth. Became established between Dunkeld and Logierait. Supposed to be at Ochtertyre as early as this; but it appears doubtful, as on Fowlis Wester, adjoining, we have no account of them before 1852. The difference of dates of arrival may, however, be accounted for by difference in age of plantations on the estates.

1847. According to the proprietor's information, first seen at Faskally, at the junction of the Tummell and Garry, in 1847. Another account gives 1857 (*q. v.*) as date of first appearance in the Tummell valley, but judging from other evidence, and the continuity of the woods between Faskally and Dunkeld, as well as the comparatively short distance from the centre, (viz. 12 miles, or 18 miles by river), the earlier date must be accepted. From that it reached up the Tummel

[1] Birds sent from Taymouth to Arran. See also under "Arran," and *antea*, under "Restoration." Arrived at various localities between Dunkeld and Logierait, where they are now resident and abundant.

valley to Bonskeid, Bohally, and Dunalastair, which are respectively 12 miles, 10 miles, and 8 miles from $\odot$.

Also appeared at Lyndoch, near Strathord (18 miles from $\odot$.) Now—1877—abundant. Area about 1000 acres.

Established at Glenalmond (14 miles from $\odot$), having bred regularly since in the "Small Glen," but almost always leaving for the more extensive and suitable woods on the south, or right, bank of the river Almond—one of the few instances in which we find the southern bank preferred. The exposure, however, lies well to the sun. Preserved. Perhaps 30 to 40 birds on the estate, taking all the season through.

1846. Is reported to have reached Gask on north, or left, bank of river Earn as early as this date; but see under 1858. Gask is 21 miles from $\odot$.

[Birds were sent to Arran direct from Sweden. Restoration in Arran partly by Taymouth birds (1843), and partly by Swedish birds. (See under headings "Arran" and "Restoration.")]

1848. This is probably nearer the correct date of the first bird shot at West Dron on the Elcho estates, though 1844 has been assigned—a female. It was shot by Mr. Robert Wood, who had it stuffed, and it is now in Lord Elcho's possession at Gosford. This must still be considered an early arrival, Taymouth being 28 miles distant, and the next earliest arrival in the district dating 1853. At present the birds are only occasional in the district, owing to the nature of the coverts.

1849. Arrived at Keillor Forest, near Methven (21 miles from $\odot$). 17 birds have been shot here in one day at Keillor Wood, and were on this occasion *all cocks*. (*Auct.* Mr. David Brown, naturalist, late of Perth, who saw them killed.)

1850. Arrived at Banff in the east of the county (27 miles from $\odot$). Area 2000 acres. Increasing very much now —1877. 25 have been shot in a day.

Appeared, but rarely, as early as this at St. Martins, near Perth (24 miles from $\odot$). For establishment, see 1860.

Arrived at Gorthie, Crieff district (16 miles from $\odot$) from Taymouth direction. Established 1856. First shot was a hybrid. After that they increased. Woods at that time about 25 years of age, and are now composed of spruce, Scotch fir, and larch; about 400 acres in extent. Of late years Capercaillies have decreased, owing to thinning of the woods to a considerable extent. Black game have now—1878—entirely disappeared, and it is believed that they have been extirpated by the Capercaillies.[1] I am indebted to Mr. Frank Norgate, of Norwich, for obtaining for me the return from this locality, and to G. R. Mercer, Esq., for kindly filling in the circular.

1852. Hybrids were obtained at Logiealmond in Glen Almond (14 miles from $\odot$), as related in the '*Zool.*' 1860, p. 7325, and birds arrived, or were first shot, at Fowlis Wester and at Monzie (18 miles and 14 miles from $\odot$). In 1859 12 were shot one day at Fowlis Wester. Not supposed to be increasing now at Fowlis Wester. Woods very extensive, joining with Abercairney, Tulchan, Balgowan, Gorthy, Cairnies, Keillor, Methven, and Glenalmond. On Monzie about 500 acres of wood. Arrived also at Balgowan—adjoining property—about 450 acres of Scots fir (see 1857 for establishment, and 1864 for increase). Also at Kinfauns, Carse of Gowrie (27 miles from $\odot$), where there are miles of pine woods adjoining. Now abundant.

1853. Are reported as being present at Strathallan as early as 1853 (21 miles from $\odot$). But see under 1857.

Became established at Woodhill, Strathardle (19 miles from $\odot$). 200 acres of wood.

Arrived in the Bridge of Earn district, at Kilgraston, Balmanno, Dron, Dumbarney, Glenearn, Pitkeathly, and part of Ochil Hills (25-28 miles from $\odot$). At these localities, it is,

[1] But see remarks under "Decrease of Black Game," p. 120.

however, for the most part sporadical—or occasionally breeding, but leaving again. The pine growth is not extensive, and is for the most part mixed with hard wood.

Arrived and became established at Dupplin Castle, north side of Earn River (22 miles from ⊙). 16 were killed here one day in 1877. This property contains about 1800 acres of pine wood. Preserved and increasing.

1854. The first appearance at Tulliallan was a hybrid, and none were seen afterwards till the introduction took place in 1856 (*q. v.*) (38 miles from ⊙). Not established here until 1864 (*q. v.*)

Arrived at Ardoch, Braco, near Auchterarder. (23 miles from ⊙). 150 acres of suitable wood. Now common and preserved.

1856. Became established about 1856 at Ardoch (*ut sup.*)

Introduced at Tulliallan, but only hens (four in number) reared, resulting in hybrids. Not fairly introduced and established till 1864 (*q. v.*)

First seen at Lanrick, Blairdrummond, and Gartencaber, near Doune. (27½ miles from ⊙). Not established till 1860 (*q. v.*)

Arrived at Condie, in Ochil Hills, on the march of Kinross (30 miles from ⊙). Probably from Invermay and Dupplin. Resident now—1877; not preserved; not increasing.

1857. Another account gives arrival at Faskally in 1857; but see under 1847.

First seen at Ballathie, near Stobhall (23 miles from ⊙). Here the suitable acreage is isolated and limited, or an earlier arrival might have been looked for.

Arrived at Ashmore and Pearsey, east of the county, in Strathardle (21-4 miles from ⊙), or earlier? *Vide* also under 1867.

First seen at Drummond Castle—*v.* '*Drummond Castle Game Book*' (17 miles from ⊙). Now abundant (see 1870).

Arrived at Blackpark, in Glenalmond (? miles from ⊙), coming from the north.

Became established at Balgowan, in the Methven and Glenalmond district, which property marches with Gorthy, Cairnies, and other properties, where there are great tracts of forest. For arrival, see under 1852 (*antea*, p. 60). For increase, see under 1864. (Balgowan is 19 miles from ⊙).

Is positively stated to have arrived for the first time at Strathallan, coming from the forests to the north (19 miles from ⊙). 2000 acres of Scots fir on Tullibardine Moor. Preserved. They increased rapidly until the number reached the limit the woods were able to support. Also became established the same season (1857). See, however, under 1853.

1858. Up to this time they were strictly preserved on the Athole properties, and scarcely any ever shot.

Arrived, and became established at Methven (20 miles from ⊙), from the direction of Birnam, and down the Tay valley by Strathord and Lyndoch. 300 acres fir. Not preserved, not increasing.

Arrived at Inchmartine, in the Carse of Gowrie (25-8 miles from ⊙). Probably from St. Martins, across the wooded Sidlaw Hills. The properties adjoin..

Said to have arrived at Strathord only this year (21 miles from ⊙), but probably earlier, as record is indefinite. Compare Lyndoch under 1847.

Arrived at Trinity Gask, between Dupplin and Strathallan, on north or left bank of river Earn, on the authority of Sir Thomas Moncrieffe, Bart., though one correspondent places it at 1846, speaking from memory, and another at 1863; but I take this date, as Sir Thomas Moncrieffe adds that his information is from memoranda in his possession. (20 miles from ⊙), probably from Lyndoch direction.

The first obtained at Loyal House, near Alyth, in the east

of the county (28 miles from ⊙), by Major J. W. Wedderburn, was a hybrid (see '*Proc. Royal Phyl. Soc.*,' vol. ii. p. 44). See also under heading "Hybrids," farther on, p. 115.

1859. Established at Trinity Gask (*ut sup.*) Area of wood on Trinity Gask 360 acres. Not increasing.

As many as twelve were shot in one day at Wester Fowlis this season (20 miles from ⊙).

About this time birds were on Abercairney, but I have no fixed date of arrival for this property. (See under 1861 and 1862.)

1860. Became established at Lanrick, Blairdrummond, and Gartincaber (see for arrival, 1856). Probably they reached this from the Drummond Castle direction. Hens, as usual, were seen first, but the remains of a cock were found also in 1856. Had it survived, probably an earlier establishment would have taken place. See "General Distribution," *infra*, p. 113.

Became established at Banff (27½ miles from ⊙). See 1850.

Arrived at Kindrogan, in Glen Brerachan, near Dunkeld (only 19 miles from ⊙). A late date, but the direction of their advance would not probably be across the hills from Dunkeld, but up the valley from Ashmore and Pearsey, in Strathardle.

Became established on St. Martins, and were plentiful by that time (see 1850). On this and estates adjoining—viz. Dunsinnan, Scone, Stobhall, etc., there are large tracts of wood. On St. Martins two woods, of 1000 and 2000 acres, are in part shared with Dunsinnan and Lord Mansfield's. They are now as abundant as black game used to be, which Mr. Macdonald Macdonald, of St. Martins, writes they have now quite driven out. Black game are now "entirely gone" from the estate. St. Martins also joins with Inchmartin estate in the Carse of Gowrie, across the wooded Sidlaw Hills.

A hybrid was obtained in Perthshire, and a nest of nine eggs found at Logiealmond ('*Zool.*,' 1860, p. 7235).

1861. First seen at Kinnoul, as far as I can gather (*v.* '*Scot. Nat.*,' vol. i. p. 114). (25 miles from ⊙).

First bird shot at Abercairney, near Crieff, on the 8th November, "but numbers were on the estate two or three years previously, but were not shot at." (James Robertson, keeper, Abercairney.)

1862. About this time very abundant at Taymouth. Estimated at 1000 birds by Lord Breadalbane, and over 2000 by James Guthrie, the keeper, but I have no doubt they reached a maximum before this.

1862. First shot at Gleneagles, near Auchterarder (25 miles from ⊙). The wood here is small, and mostly mixed hard wood. Had Gleneagles been well planted with pine it would have formed a natural avenue of advance into Glendevon, and across the Ochils into Kinross and West Fife.

First shot at Stobhall (23 miles from ⊙). '*Drummond Castle Game Book.*' First seen at Moncrieffe—a female, October 27. (Memorandum by Sir Thomas Moncrieffe.)

Established about this date on Abercairney, near Crieff (19 miles from ⊙). Now as many as 12 seen in a day. Preserved; increasing slowly.

1862. About this date, "16 years ago," one was shot on Glengyle Hill (?), near Loch Katrine (28 miles from ⊙). This date is a long way in advance of others in the neighbourhood, and I fancy there must be some mistake.

1863. For Gask, see 1858.

Said to be becoming scarcer at Murthly. Were much more numerous 15-20 years ago than now. One reason assigned is, that the late Mr. Condie killed them down for a few years in Rohalzion woods opposite. They are now again increasing.

Established at Moncrieffe (26½ miles from ⊙). Now 50

or 60 seen in a day. Preserved. 500 acres and upwards. Not increasing. Sir Thomas Moncrieffe notices that a number migrate every year.

See Abercairney, Crieff district, under 1861 and 1862; and map.

1864. Established at Tulliallan, an introduction by eggs, procured by Lord Balfour of Burleigh from Freeland. For account of introduction, see under Chap. xx. p. 116. Now very abundant at Fowlis Wester.

Bred at Aberuthven wood, south side of river Earn, east of Auchterarder—an isolated clump of about 60 acres of fir (24 miles from ⊙), and continued to do so sporadically, until 1869 or so, when they left altogether, the wood probably becoming too old. In 1861, I shot in this wood, and it was then of pretty old growth.

First seen at Kippenross (27½ miles from ⊙). Woods on Sheriffmuir pretty extensive.

Arrived at Balmyle, in Strathardle, and at Black Craig, Strathardle, "probably from Athole" (?) (21 and 20 miles from ⊙ respectively). About 40 acres of wood on Balmyle estate and 100 on Black Craig. Not abundant; preserved; increasing.

Arrived at Glentarchy and Ayton, in the Bridge of Earn district (30 miles from ⊙). See 1856, where arrival is recorded at Condie, in the Ochils, still farther to the south. Discrepancy may be accounted for in the *later* growth of wood at Glentarchy.

At Balgowan the keeper saw 100 birds in one place at one time in 1864 (see 1852 and 1857), and 13 have been shot in one day.

1865. First seen at Braes of Doune—Lord Moray's property (26 miles from ⊙). Apparently a discrepancy, but various causes must be taken into consideration. They appear to have arrived at Lanrick and Gartencaber, coming from

Ardoch and Drummond Castle, before populating Braes of Doune. But even now (1877) they are only uncertain in the woods there, and on Argaty adjoining. The woods are somewhat isolated, and not very extensive.

Appeared at Luib, Glendochart (25 miles from $\odot$).

First seen at Pitcairlie, Fife march (33 miles from $\odot$). See under Fife, 1865.

Established at Monzie, Crieff district (15 miles from $\odot$); Ferntower (15 miles from $\odot$); Ochtertyre (15 miles from $\odot$). These are now quite parts of the stronghold of the species in the south of Perth.

Established at Dalnaglar, Glen Shee, east of county (25 miles from $\odot$). For full notes on this locality, see under "Decrease of Black Game," etc.

Established at Tulliallan. See under "Hybrids" and "Restoration."

Arrived on Kincardine estate between this date and 1860, probably from Strathallan or Drummond Castle. Kincardine, near Auchterarder (24 miles from $\odot$).

1866. First seen at Aldie, near Crook of Devon, march of county Kinross (32 miles from $\odot$), and established same year. Not preserved formerly, but present tenant does preserve. At present rare, but likely to increase.

Established at Glentarchy and Ayton (see 1864). Not likely to become numerous. Glentarchy is really in Fife, *q. v.* under 1866.

First seen at Invermay (26 miles from $\odot$), near Dunning. Occasional at present (1877). Common in autumn; eight seen in a day. About 200 acres mixed wood. Increasing, but do not breed here.

1867. First shot at Airthrey, near Stirling, south of county (31 miles from $\odot$). Five seen at Bridge of Allan, Christmas 1877.

[Said to be established only in 1867, at Ashmore and

Persey, east of county, but the other record is best authenticated. See under 1857.]

Very abundant at Ochtertyre (see 1865).

1868. A female first seen at Cardross (32 miles from $\odot$), S.W. of county (see under 1870).

Arrived at Garvock from Gask or Dupplin. Garvock is near Dunning (27 miles from $\odot$).

1869. Abundant at Dunsinnan, to the N.E. of Perth (27 miles from $\odot$). I saw at least thirty there one day when shooting in December. I have not received dates of arrival or establishment at this locality.

1870. First male seen at Cardross, near Lake of Menteith (see 1868 and 1871).

Abundant at Drummond Castle, Earn valley, near Crieff. This is now a great stronghold of the birds in the south of the county (17 miles from $\odot$); strictly preserved now; increasing very much still. The adjoining estates are Strowan, Ardoch, Strathallan, Abercairney, etc., upon all of which there are considerable areas of suitable and continuous woods. Torlum[1] Hill, the highest in Scotland which is wooded to the summit (by last survey 1291 feet), is on the march between the Drummond estates and Strowan, and is considered to be more thickly populated by Capercaillies than any other localities in the Crieff-Comrie district.

1871. Established at Cardross. (This is a good illustration of the succession and arrival of female and male, and establishment—see farther on, p. 113.)

Arrived at Stronvar, Balquhidder. (For attempt at introduction see 1845). Only occasional at present, and rare (only 24 miles from $\odot$, but across hills, in a direct line).

[1] *Torlum*, from Gaelic *Tor-lom*, or *The bare hill*. "This appearance is now changed, it having been planted in the end of the last century; but drawings which exist of Drummond Castle show that the name was truly descriptive (*vide* Robertson's '*Gaelic Topography of Scotland, and what it proves*,' 1869, p. 457).

Probably advanced from Comrie and Crieff along Lochearn-side, or possibly from the Doune district, or possibly from Glen Dochart. The Doune district route is the least likely.

1872. Arrived at Leny, near Callander (25 miles from ⊙). Probably from Lanrick and Doune (see 1874).

Arrived and bred at Ashintully, 15 miles north of Blairgowrie (14 miles from ⊙).

1873. Only occasional at Balquhidder still (24 miles from ⊙). Near Stronvar. Up to present time (1878) only occasional.

1874. Twenty-three shot one day at Dunkeld in 1874 (see 1843-4).

Established at Leny. Ten to twelve birds on estate now (see 1877). In Glenartney, behind the hill from Leny, there are no Capercaillies. It is probable, therefore, they advanced to Leny *vîa* Doune and Lanrick.

1876. First shot in Dearry woods, 5 miles from Lochearnhead (about 15 miles from ⊙, across the hills).

Very abundant at Moncrieffe; thirty seen in a day (see 1863).

A young retriever dog caught a young chick at Leny at the end of July. This is the first positive evidence of their having bred here.

1877. First seen at Kippendavie, near Stirling (28 miles from ⊙); woods scattered.

First seen at Evelick, Carse of Gowrie (24 miles from ⊙).

First seen at Ardvorlich, south side of Loch Earn (18 miles from ⊙). Patch of isolated larch wood of considerable age.

First seen at Loch Kennord, between Tay and Strathbraan district. Probably came up the valley from Strathbraan, having reached round from Taymouth almost in a circle; or possibly, as wood was planted higher up the slope of the hill on the north side of the Tay, or grew to a greater height, the birds may only at this late date have attained an altitude sufficient to permit them to see the wooded banks of

Loch Kennord, at the back of the lower hills which are on the south side of the Tay. While all the way down the river from Dull, on the north bank, to Pitnacree, near Dunkeld, the birds occur, it is somewhat strange how *almost* unknown they are across the river at any locality below Taymouth; the want of wood, however, sufficiently accounting for it. Increase in the growth of the trees at Loch Kennord also, of course, may have been the cause of attracting them. I think there can be little doubt that the bird's power of vision is great, and that this is a powerful factor in their distribution. They are often seen plunging from the wooded heights of Moncrieffe Hill, and making for the woods which lie scattered over the eastern spur of the Ochils, and which are situated upon the properties of Invermay, Condie, Kilgraston, and the neighbourhood of the Bridge of Earn—a distance of four or five miles at one flight. Though never occurring on Seggieden property, they are established close to it on Kinnoul, and Colonel Drummond Hay has often seen them passing high overhead, making for the wooded slopes of Evelick in the Carse of Gowrie.

1877. Had reached Tyndrum, at head of Glendochart, only 27 miles from ⊙, but locality far removed from suitable woods (see remarks farther on, under "General Remarks," p. 105).

1878. At the present time Capercaillie are abundant at Dall, on Lochrannochside, as I am informed by Mr. George Galbraith, Skye, who knows the district well between that and Faskally, along the Tummel valley. He writes (*in lit.*):—"At Dall they seem to be more numerous than in any other place—[*i.e.* on the Tummel above Faskally.—J. A. H. B.] The Black Wood of Rannoch seems to suit them. I have seldom or ever gone trout-fishing on Rannoch without seeing several."

They are resident. Two males were first observed in the Black Wood, and lived for two or three years in celibacy. A female was then introduced by the late Struan Robertson,

Esq., and these three birds were the ancestors of all the Capercaillie now in the Black Wood, as I am informed by another correspondent resident on the spot.

1878. At Glen Queich (about 5 miles from ⊙), it has again been observed. Here there is only a small bit of young plantation and a few scattered large trees.

Beyond Perthshire limits, the Capercaillie had reached Forfarshire in 1859, and farthest in that county at Stracathro, near Brechin, in 1865 (54 miles from ⊙).

In Fife it was first seen in 1863, and reached farthest in that county in 1875, at Kinaldy (50 miles from ⊙).

In west of Fife about 1871, but these birds probably came from the separate ⊙ at Tulliallan, and reached Pitreavie in 1877, near Dunfermline (43 miles from ⊙).

In Stirlingshire first appearance was in 1856, probably a strayed bird from Tulliallan,—a female shot on an open common at Stenhousemuir, near Larbert (38 miles from ⊙). The next was about 1866, at Gardén, near Kippen (35 miles from ⊙). Thus it is seen they arrived here before they arrived at Cardross (1868). But this is accounted for by the fact that Gardén woods, which are frequented by them, stand on a height, and are visible from Doune, whence, doubtless, they came, whilst Cardross woods lie in a hollow, with rising ground intervening. They reached their farthest in Stirlingshire, at Dougalstone, Milngavie, near Glasgow (45 miles from ⊙), in 1877, where they bred for the first time.

In outlying counties,—they reached Linlithgow in 1872; Edinburgh in 1876 and 1877, the latter being probably one of two birds seen shortly before at Pitreavie in Fife (*vide* map). Towards Loch Lomond, an occasional bird had been got as early as 1867, and one in 1878, at Ross Priory. In Argyleshire at the Black Mount (an unsuccessful introduction).

For *particulars* of *advance*, I refer readers to separate accounts of the counties. We now proceed to Forfarshire.

CHAPTER VIII.

EXTENSION IN FORFARSHIRE.

THIS county ranks next to Perthshire in its Capercaillie population, and, notwithstanding that the proprietors look upon the Capercaillie as rather a mischievous bird, it has succeeded in gaining good foot-hold. They are not, however, preserved in Forfar to the same extent as in Perthshire, and on several estates they are killed at all opportunities. I still require, I believe, further data from the northern parts of Forfar, especially Glens Clova, Prosen, Upper Isla, and North Esk.[1] Avenues of advance into Aberdeenshire may occur in Glens Clova and North Esk, but on this head I will speak later.

1856. The earliest appearance of the birds in Forfarshire appears to have been in 1856. Arrived at Lindertis, near Kirriemuir (about 35 miles from ⨀). Resident. In 1868 six brace were shot one day. There are 800 acres of Scots fir and blaeberry. "The cover in which we mostly find them is about 45 years of age." They do not increase here, being too much disturbed; and being close to some of Lord Airlie's large wooded hills, the birds go there, where they are not disturbed. "There are about the same number as on arrival." [This points to an emigration of a large body of

[1] The farthest north they go in Forfarshire is to Shelgreen and Benscreavie woods, about ten miles north of Kirriemuir. At Glenhead, twelve miles north of Alyth, they have also appeared, but orders have been given to shoot them down "on account of their destructiveness to black game."

the birds at once.] Black game are now decreasing, "but not on account of Capercaillies."

1859. A male bird was shot on the 5th January 1859 in the united covers on the four estates of Aucharroch, Ascreavie, Kinchine, and Airlie, on which the wood is about two miles long by three-quarter mile wide, and consists of spruce, pine, and larch. "The birds shift, accordingly, as they are disturbed." Probably arrived here from Bamff by Alyth, in Perthshire. "Hens are most numerous." *Auct.* Mr. Thomas Phillips, twenty years gamekeeper at Ascreavie (32 miles to 34 miles from ⊙).

1860. I have one record of the bird in Forfarshire as early as 1860, but this must have been a thoroughly strayed bird, to reach as far as Easter Ogle (43 miles from ⊙), in Glen Esk.

A probable establishment took place at Airlie and Ascreavie, as establishment is found to take place usually immediately after the arrival of the male bird (see 1856 and 1859).

1862. The birds do not appear to have become established in south of Forfarshire before 1862, when they arrived and became resident at Fotheringham and Tealing (the properties of W. Scott Elliott, Esq.) The direction of their advance was probably from the N.E. spurs of the low-wooded range of the Sidlaw hills, and from the flatter country lying between these hills and the rivers Tay and Isla, as we find dates of arrival and establishment in the adjoining parts of Perthshire, as well as suitable country favouring this line. Fotheringham and Tealing are respectively 38 miles and 39 miles from Taymouth ⊙; and there are about 2500 acres of suitable wood upon the two estates.

The same year an impulse was given by a successful introduction at Cortachy by birds, where of late 16 have been seen in a day, and 8 shot. Not preserved. Increasing. Black game decreasing.

1863. In this year two male birds were shot at Kinnordy, and since then they have been resident, having plenty of scope to increase their numbers in large fir and larch woods, of between 400 and 500 acres in extent. These birds probably arrived from the direction of the Bamff estate on the borders of Perth, or it is also possible that they came northward from the Sidlaw hills, as we find them also at Glamis in 1863, though not established there till 1865 or 1866. The former advance, we think, is the more probable of the two, as correspondents agree that the first place visited by the birds in this part of Forfar was on Lord Airlie's grounds, close to the county march. I do not, unfortunately, hold an exact date for their arrival at Airlie Castle, but I have a return from the united estates of Airlie, Ascreavie, Aucharroch, and Kinchine, which is sufficiently exact.

1864. Occasional birds appeared at Kincaldrum, probably wandered from Airlie Castle or Kinnordy. The woods are about 100 acres in extent, but the birds have never become resident.

1865. Arrived at Fearn; becoming established in 1868 (*q.v.*) (44 miles from $\odot$). The area of wood is not great, only some 40 acres, but this appears to be sufficient to hold them.

It was in this year also that Capercaillie reached Stracathro, the farthest point in Forfar, in the Brechin district. This is still the only locality in the immediate vicinity of Stracathro where they are resident and breeding, though stragglers are shot on most of the adjoining estates. In 1877 there were 3 or 4 broods, and my friend Mr. W. Horn saw 7 or 8 birds in one day's cover-shooting, though the proprietors do not protect, but try to keep their numbers from increasing. Stracathro is 54 miles from $\odot$, and the direction of their advance to this point is almost in a straight line from Dunkeld. At Dalnaglar, a point in Glen Shee almost upon the Forfar and

Perth march, they also arrived in 1865; and this illustrates remarkably well, amongst many other statistics, the preference for following valleys, and their objection to crossing mountains, Dalnaglar being only 25 miles from $\odot$ in a straight line.

1866. Arrived about this date, or a year or two earlier, at Auchterhouse, a locality considerably nearer to the Sidlaw hills than Tealing or Fotheringham. My local topographical knowledge of this part of the country is not sufficient to allow me to offer a reason for this apparent discrepancy, unless it be found in the fact of the then age of the timber on the different estates, which was ready for cutting in 1877 on Auchterhouse. A large quantity was cut in 1877, and a consequent decrease in the numbers of Capercaillies is looked for. If all old wood about 1866, and Tealing had younger wood, the birds might pass over to the latter.

Became established at Easter Ogle, where it is said to have appeared as early as 1860, and is now common, as many as 20 to 24 having been shot in a day, although the acreage of wood is only from 100 to 150 on Easter Ogle and Deuchar, principally pine and larch. They are preserved, but are not now increasing.

1868. Became established at Fearn, where they are not numerous, the acreage being small (see 1865).

Arrived about 1868 at Glen Ogle, or possibly a little earlier, where they are not now increasing. Mr. James Robbie, head-keeper, has observed their tendency to migrate.

Arrived at Panmure (45 miles from $\odot$), where, however, it is still only occasional, or where one or two may be seen in a day. Came from Fotheringham.

1870. First appeared—a male—at Balnamoon, in the Stracathro district, north-east of the county. Now resident; preserved, and increasing in a fir wood of 300 acres (50 miles from $\odot$). Occur also on Hunthill, Careston, Glen Ogle, etc.

JUNCTION OF FEUGH & DEE.

Arrived about 1870 at Ballintore, from Airlie, and established about 1872, in woods of about 80 acres, where they are preserved, but are not increasing at present. Occur also on neighbouring properties of Aucharroch and Ascreavie.

1871. Arrived from the west about 1871 at Auchnacree (32 miles from ⊙). Established about 1873. Not preserved, but still increasing, in a wood of 100 acres. As usual, in most cases, a female was first observed.

Arrived about 1871 at Careston, Glen Esk (46 miles from ⊙). Not preserved, but still increasing. Acreage 150 of pine, and 90 of birch, etc. Probably came from Fearn or Balhall.

1877. Are stated to reach along the south slope of the Grampians, and may be expected to spread into Deeside, in Aberdeen, and through Kincardineshire, probably *viâ* Glendye, which runs into the Feugh, and so joins the Dee at Banchory. Glen Dye is wooded far up into the dividing range of hills, but there is a considerable portion of moorland, which would require to be bridged by the birds' flight. Still this appears to me to be the probable outlet beyond the Grampians.[1]

Capercaillies occur also in Forfarshire at various other localities, from which, however, I have no dates of arrival or establishment. Amongst these I may mention :—

Kinnaird, near Brechin ; Balnaboth, in Glen Prosen, a far north locality ; Lindertis, near Kirriemuir ; Noranside ; Balhall.

[1] *Vide* Kincardineshire. Since the above was written they have followed this route.

CHAPTER IX.

EXTENSION IN FIFESHIRE.[1]

PERMANENTLY inhabited by the species, Fife can, however, hardly claim to be distinctively a Capercaillie country, as the woods are not, as a rule, of very great extent, as in Perth and Forfar. At one or two localities, however, they are increasing steadily in numbers, as, for instance, at Lathirsk, east of Auchtermuchty, where they were introduced a few years ago, and where, in 1877-8, there would be from 20 to 30 birds.

1863. The first Capercaillie seen in Fife, so far as I can learn, was shot fifteen or sixteen years ago at Rankeillor, to the west of Cupar (37 miles from ⊙), after a long continued gale from the north—*auct.* Robert Tullis, Esq. It is true, Captain H. W. Feilden informs me that he has an indistinct recollection of a Capercaillie having been killed at Falkland Castle, where there are extensive fir woods, as early as 1847, but I have been unable to obtain more exact statistics.

1864. Stragglers appeared about Colessie, near Auchtermuchty, and around that neighbourhood, and at several localities along the Fife-Perth march, but it was not until much later that the birds appeared to take a hold on the county.

1868. A great blank now occurs, and we hear nothing more of the Capercaillie in Fife until 1868, when the birds arrived

[1] I find I have returns from at least twenty localities in Fife.

at Pitlour (32 miles from ⊙), near Auchtermuchty, probably coming from the direction of the Bridge of Earn district, where, as has been shown under Perthshire, the woods are not extensive, and are greatly mixed with hard wood. Nor did establishment take place at Pitlour or in Fife until 1871 or 1872.

1871-2. Established at Pitlour as above stated, and in 1873 from eight to ten were seen in a day, and six were shot in a season. They are not preserved, but are thought to be increasing in a wood of some 250 acres.

About the same time stray birds appeared at Nether Kinneddar, in the south-west corner of the county (37 miles from ⊙, these having probably strayed from the separately-introduced stock at Tulliallan centre, in the detached part of Perthshire, near Kincardine-on-Forth, as mentioned under the latter county. They have never become fairly established here; although birds remain during the breeding season they do not stay throughout the year, but repair to larger covers on West Grange, Tulliallan, and Brucefield, the woods on Nether Kinneddar being, as yet, too young, and not of great extent—*auct.* W. C. C. Erskine, Esq.

1873. Two birds appeared at Pitfirrane, near Dunfermline (40 miles from ⊙), where there are some 200 acres of old fir and mixed wood. At present they are seen here every spring, but do not remain. An attempt was made to introduce them by eggs from Taymouth, but an unusually wet season (1877) killed the young birds—*auct.* Mr. Thompson, head gamekeeper, Pitfirrane.

1874. An introduction, which eventually proved successful, took place at Lathirsk (37 miles from ⊙), near Falkland, at which latter place, however, they are said to have appeared before they were introduced at Lathirsk. The introduction was from Monzie, in Perthshire.

1875. In the spring of 1874 or 1875 a male bird was seen at Lathockar, in the east of Fife, supposed to have wandered from Falkland—the nearest place to Lathockar, where they breed, being 20 miles off—Lathockar is 50 miles from ⊙ at Taymouth, and about 38 miles from ⊙ at Tulliallan; but from the nature of the intervening country it is improbable that they came from the latter. There are only some 50 or 60 acres of suitable wood on Lathockar. Possibly also it may have come from Lathirsk, 15 miles off.

1877. The Capercaillie in 1877 appeared at quite a number of localities in Fife, but from the insufficient acreage or unsuitable character of most of the covers, I believe their attempts at advance in this direction cannot be successful in the immediate present. They appeared at Pitreavie, near Dunfermline. Mr. Paterson of Dalnaglar, in Perthshire, lessee of Pitreavie, writes that he saw two birds there about the last week in November 1877, and a few days afterwards Lord Rosebery shot a female at Dalmeny on the opposite side of the Firth of Forth, in all probability one of the same birds seen at Pitreavie, *showing* that the birds *will* cross water *if* they can distinguish suitable forest land on the other side, as I have already mentioned in my concluding remarks under Perthshire.

Other localities, where they appear for the first time in 1877 in Fife, are Auchtermuchty, Markinch—female seen—Leven—female shot—Pitliver, and Scotscraig, and others.

In the south-west part of Fife, as plantations increase in size and number, we may look for an advance from the somewhat isolated and fully-stocked districts of Tulliallan, West Grange, and Brucefield, but scarcely before then.

Had Glen Eagles, in the Ochils, been a forest-clad glen, we might, with some certainty, have looked for an advance through it from Perthshire.

CHAPTER X.

EXTENSION IN KINROSS-SHIRE.[1]

THERE are no extensive pine woods in Kinross. Consequently, the Capercaillie is a rare bird in the county, comparatively speaking. The only locality where they are resident in the county is at Cleish Castle. Aldie is really in Perthshire, near the Crook of Devon (*v.* Perthshire under 1866). As there are considerable woods upon Aldie, and the birds are likely to increase there now, Kinross will probably be supplied with occasional stragglers, but until more ground is planted it is not likely to become resident in any numbers.

1866. This is the earliest date I have for their appearance in Kinross—probably a stray bird from Aldie. This was at Craigow, Milnathort, where, at the present time, they only appear occasionally in autumn and winter, leaving in spring. The woods are small and old. Perhaps altogether two pairs may be seen in a season.

A few birds have been seen from time to time in the Knock Wood in the Ochils. In 1873 one was seen at Thomanean, side of Loch Leven. It occurs occasionally at Portmoake, in the northern and eastern portions of the county, probably coming from the Perthshire side of the Ochils and from Fife. They have also been shot at Touchie,

[1] My returns from Kinross are 12 in number, representing all the important estates in the county.

Craigfarg, and Athronhall, where five have been shot from time to time.

Beyond the above there seem to be few—if any—statistics forthcoming from Kinross, and Mr. R. Burns Begg says I may rest assured that I have received all that is obtainable, from Messrs. Bethune and Henderson, to whom I am indebted for almost all my notes from this county, through the introduction of David Syme, Esq., sheriff of Kinross.

CHAPTER XI.

EXTENSION IN CLACKMANNANSHIRE.

ALTHOUGH marching with a district—Tulliallan, Brucefield, and West Grange, in the south of Perthshire—thickly populated by Capercaillies, Clackmannan cannot claim any appreciable share of them. There are few large woods suitable, although it is almost surrounded by them. Thus only occasional birds are seen, and these rarely. Two female birds were shot at Mixed Park about 1875. It has not occurred upon Kennet, the plantations not being large enough. Lord Balfour of Burleigh, however, took an active part in introducing them to Tulliallan.

Mr. Robert Gray, in "Ornithological Notes" in the '*Proc. Berwick Nat. Club*,' 1877, p. 354, mentions them as found in the woods of this county "in considerable numbers," but this can refer only to the Earl of Zetland's woods.

CHAPTER XII.[1]

EXTENSION IN STIRLINGSHIRE.

STIRLINGSHIRE, there can be little doubt, forms a natural link in the chain which will in time lead the Capercaillie into the southern counties of Scotland. The avenues of advance into the county are probably two in number; one leading from the separate introduction at Tulliallan in 1856 and 1864 (*vide* Perthshire); and the other, and perhaps less isolated, and more generally adopted route, from the south of Perthshire—from Stirling westwards. The largest "*waves of distribution*" seem to have pressed through the narrow pass near Stirling, rounding the spur of the Gargunnock hills, and flooding the hill slopes and wooded uplands, reaching as far, at the present time (1877), as Chasefield covers, near Denny, which are forty miles from Taymouth; or even farther, as, in 1877, Capercaillies bred for the first time at Dougalston, Milngavie, near Glasgow (46 miles from ⊙), certainly the farthest south-westerly point at which we can record the arrival of birds bred from the original Taymouth stock. Their next steps of advance southward, in all probability, will be to the extensive woods of Callendar, near Falkirk, and thence eastward and southward. As mentioned under "Out-

[1] From Stirlingshire I have 32 returns, which must fairly represent all the estates of importance in the county which are suitable for the safe harbourage of the species.

lying Counties," p. 88, it will be seen that already there are indications of advance in this eastward direction in Linlithgow and Edinburgh, though no doubt *stream-waves* from Stirlingshire and from Tulliallan centre will coalesce, as partially exemplified by the Dalmeny example, shot in 1877, which, as already shown, in all probability came "across seas" from Pitreavie (*v.* Fife).

Another outlet into Stirlingshire from Perthshire may possibly be found in the woods of Cardross, at the side of the Lake of Monteith; but there is this against that line, that Cardross woods lie in a hollow, and at present—except isolated woods at Gartmore and Duchray (near Loch Ard), and on Garden, and stripes of wood on Ross Priory, Loch Lomond side—there is little to lead them from it in a direction round the western spur of the central range of hills in Stirlingshire. It is perhaps uncertain at present whence the birds which bred at Milngavie came; whether by this route, or from the eastward by Chasefield, near Denny; but we think *probably* from Cardross direction, as we find connecting links in Stirlingshire of occasional occurrences at Ross Priory, Killearn, and Culcreuch, near Fintry. When the woods grow higher upon Ballikinrain estate, near Fintry, another natural '*forest stepping-stone*' will be afforded, but that cannot well take place for a number of years yet.

The principal distribution of wood in the county is on Meiklewood, near Stirling, and westward on Leckie and Gargunnock, to Boquhan, Arngomery, and Garden, along the north slope of the Gargunnock and Fintry hills. There are young plantations of considerable extent upon Ballikinrain. Again, southward from Stirling, is a range of wood from Meiklewood, through Touch, Sauchie, Auchinbowie, Daleswood, Torwood, Quarter (where 60 acres of wood is just 34 years planted), and Chasefield—all more or less occupied at present by the birds. These latter properties and localities lie along the eastern slope of the Touch and Denny hills, a part of the great central mass of hills which extend around the sources of the Carron and Endrick rivers.

1856. The earliest record of the occurrence of the species in this county was a bird—a female—shot upon Stenhousemuir, and for a long time afterwards in possession of Mr. Caddell of the Grange, Linlithgow. Stenhousemuir is an open, flat, grass common, used annually for the great autumn cattle and sheep markets, usually known by the name of "*The Falkirk Trysts.*" This bird, I believe, could only have come from Tulliallan, and must be looked upon as a very exceptional occurrence (compare under Perthshire, 1856). The next nearest point was the Doune district in Perth, but *there* the birds only appeared for the first time this same year.

1863. In 1863, birds bred at Dunmore, which is only separated from Tulliallan by a narrow part of the river Forth. The Dunmore and Airth woods are not extensive, otherwise we might certainly have expected a larger irruption from Tulliallan by this channel. These eggs, however, were not hatched out. [*Vide* '*Zool.*' 1867, p. 897.]

1866. First observed at Garden, and bred there in the "Kippen Firs"—a wood close to Loch Laggan (35 miles from ⊙). These birds probably came across the valley of the Forth, direct from the Doune district, from the wooded estates of Lanrick, Blair-Drummond, and Gartencaber. The Kippen Firs and neighbouring woods of Arngomery would be visible to the birds from Doune.

1867. Appeared simultaneously on Torwood Estate and on Denovan, near Larbert and Denny (both about 38 miles from ⊙). On these estates together, and on Quarter—also adjoining—there is a considerable acreage of suitable wood, and they have become firmly established.

In 1867-68 a female was shot at Ross Priory by Sir George Leith-Buchanan, Bart., who did not again meet with a bird there until 1877-78 (31st December 1877), when he again shot a female bird.

In 1868 Capercaillies bred in Torwood for the first time. "Three or four birds are seen frequently by the keeper, and it may be considered as established, as protection is afforded to these birds, and Torwood is in every way suitable for their increase. There can be little doubt they will increase, though perhaps by degrees and slowly." ['*Extract from old Note Book on Birds of Stirlingshire,*' 1868.—J. A. H. B.]

1869. On Arngomery, close to Garden, the first—a female—was shot also in the "Firs of Kippen" before mentioned.

1870. First observed on Sauchie property, near Stirling, and became established (35 miles from ⊙), probably from the Stirling direction, but also quite possibly from Tulliallan, as birds might be led there through the woods of Dunmore, Polmaise, and Carnock, in the Carse of Stirling. Now (1877) there are about 40 or 50 birds on Sauchie. The first birds seen were two females and one male, and the two hens had nests close to the avenue, and within 100 yards of the house.

Became established at Torwood, coming doubtless from Sauchie or from Tulliallan direct, or through Sauchie from the Stirling direction.[1]

A female was seen several times on Dunipace by myself, and was afterwards shot on Denovan (38 miles from ⊙).

1871. The first seen at Boquhan—a female. Then no more till 1875, when a hen and brood (35 miles from ⊙). Either come from Garden, marching with it, or direct from Lanrick. Amount of wood limited. I have shot over both Boquhan and Garden, and do not think the birds are ever likely to become numerous.

A female not unfrequently seen on Larbert House grounds in the autumn, marching on the east with Dunipace; but the acreage of wood is limited, though possibly large enough to hold a few birds if not much disturbed. There is more of spruce than pine in the covers.

[1] In April I myself saw a female in Torwood. ('*Note Book,*' 1870-71.)

1874. One stray bird was shot at Rowardennan, on Loch Lomond, about this date.

One—a female—was shot at Quarter, near Denny, in the summer, and another in October of the same year. It is now established here in a wood of 60 acres of spruce, larch, and Scotch fir, which is just 34 years old. Birds here probably arrived from Torwood on the east, or from Auchenbowie and Sauchie on the north. Auchenbowie held birds at this time, though they did not become resident till the year after—1875.

1875. Birds became established "not prior to 1875," on Duchray, near Loch Ard, just on the border of Perth and Stirling, and 33 miles from ⊙. These birds no doubt came from Cardross, through Gartmore (where, however, they are only as yet occasional). There are about 200 to 300 acres of suitable wood on Duchray and adjoining property, of which 100 to 150 are on Duchray.[1]

Occasionally seen on Carbrook, near Larbert, marching with Torwood; but the covers are mixed, and not quite suitable.

Occasional on Chasefield, when only hens were seen in this year and in 1876 (40 miles from ⊙).

Became resident on Auchenbowie, marching on the north with Sauchie.

Bred for the first time at Boquhan. In 1876 eight birds were seen, but the keeper at Arngomery shot four or five hens, and they did not breed in 1877.

1876. Two hens seen on Drum farm, Earl's Burn, up the Carron valley above Denny, in an isolated patch of firs in the midst of moorland (37 miles from ⊙). These birds, no doubt, came from Quarter woods, being attracted by the view of old Scotch fir on the horizon at Sheppytroutie, whence again they could see the younger plantations on the hill slopes at the

[1] The lessee of the Duchray shootings has never himself seen any Capercaillies on the ground.

Earl's Burn. There is not sufficiency of cover here to hold them permanently.

1877. Nine birds have been seen in all on Airthrey, but none remain from March to August.

1877. Males first seen in Chasefield (40 miles from ⊙).

1877-78. Occasional on Gartmore within the last few years (34 miles from ⊙).

A male seen at Carbeth, near Killearn, about 13th November 1877, and continued about till January 1878 (40 miles from ⊙).

Bred at Dougalston, Milngavie (46 miles from ⊙).

Resident at Quarter (37 miles from ⊙). Wood 34 years old.

Five seen at Bridge of Allan, 4th January 1878 (28 miles from ⊙).

Only occasional still at Dunmore (35 miles from ⊙).

Stray birds appeared on Carbrook, which, however, marches with Torwood, whence, doubtless, they came. As both sexes have been seen, sporadical or possibly permanent residence may here take place; but the woods are not extensive, nor very suitable.

In May 1878 I again saw a female on Dunipace, flushing it from the ground, amongst blackberry bushes, in oak coppice-wood.

1878. A bird was killed upon Callendar estates about the middle of November this year, the first seen. As already noted, it would probably come from either the Chasefield coverts or from Torwood.

We may expect their establishment here shortly. Woods extensive and suitable.

At Buchanan, near Loch Lomond, a pair bred (*auct.* J. Lumsden, *fide* Sir George Leith-Buchanan, Bart.)

CHAPTER XIII.

EXTENSION IN OUTLYING COUNTIES.

LINLITHGOW.

LINLITHGOW, marching with Stirlingshire, will, in course of time, probably receive its supplies from the latter county, through Callendar woods, near Falkirk, where the species has as yet only once been met with (1878), although these extensive woods, and others on Bonnymuir, are within sight of Chasefield and Torwood covers. It is also a possible line of advance, which has already been indicated by the advent of the female bird shot at Dalmeny Park, near Queensferry (48 miles from ⊙), by Lord Rosebery; which bird was no doubt one of the two seen a few days before by Mr. Paterson at Pitreavie, near Dunfermline.[1] But this line of approach is less likely, as it will be dependent upon the stocking of the woods of Pitreavie and Pitfirrane, on both of which estates the wood is rather old, or of too mixed a character, to afford good harbourage or make an increase likely. The advance of the Capercaillie from this direction will probably be dependent upon planting on the Fife side of the Firth.

[1] Since this was written I have been informed by the keeper at Dalmeny Park that another female bird had been seen at Dalmeny about six years before, say 1871; and another has been seen on Craigie Hall since Lord Rosebery shot the above.

In the account of the Capercaillie prior to extinction, I have already noticed the birds shot by Captain Stanton near Bo'ness in 1811.

In 1872 a male was seen by Captain Maynard in Kettlestone plantations (44 miles from ⊙, and 5 miles from Tulliallan, *across sea*).

MID-LOTHIAN.

1876. One bird was distinctly identified in Mansion House Road, at the Grange, on the outskirts of Edinburgh, in May 1876. The bird—a female—flew past the gentleman who identified it within 20 yards. It came from the south, and flew away due north, heading across the Meadows, directly for the space between St. Giles and the Tron Church towers. Possibly this might be the bird shot at Dalmeny in November 1877, but I doubt if it would have remained there so long without being detected. My idea is that, wherever it came from, it headed for the church spires, mistaking them at the distance for tops of pine trees on the horizon. Edinburgh is 54 miles from Taymouth ⊙, and 18 miles in a direct line from Tulliallan ⊙. It is rather a curious instance of the extreme wandering propensities of the species. It is reported as having come from the south. I have no statistics from any localities to the south of Edinburgh to make me suppose that its origin was there. It must have been, I fancy, a truly wandered bird, like the earlier records in Fife and Stirlingshire.

DUMBARTON.

Birds have once or twice been shot in the Loch Lomond district, probably strayed birds from the direction of Cardross. They have been obtained at Ross Priory by Sir George Leith-Buchanan, Bart. (see under Stirlingshire), and on the islands of Loch Lomond ('*Proc. Nat. Hist. Soc.*,' Glasgow, vol. iii. p. 68); also in Stirlingshire.

The only one ever heard of actually within the confines of the county, was flushed at Kearnine (?) Wood, at the higher end of Loch Lomond—as Mr. John Colquhoun informs me—but he adds that he is rather sceptical about it. Mr. John Colquhoun offered to introduce them on Rossdhu estate, but his brother—the late Sir James Colquhoun, Bart.—did not wish to encourage them, as he considered that they give little sport, and are overbearing to all other game within their reach. If this record is correct—and I must say I see no reason why it should not—the bird probably found its way from that point on the west slope of Ben Laoigh, whence they found their way also into Argyleshire, at Ardkinglass (*q. v.*) Either route is open to pioneer birds from the said shoulder of Ben Laoigh, though that towards Glen Fyne to Ardkinglass, may, prove—and apparently has proved—the more enticing of the two, as will be seen under Argyleshire.

ARGYLESHIRE.

Of its former occurrence in the county we have the somewhat vague record in the '*New Statistical Account*' of Ardchattan, where it is said, " Nor have we reason to doubt that the stately Capercaillie once rejoiced amongst the fir woods of Glen Etive (*op. cit.* p. 481).

Besides the above, we have indications of its having lingered in this part of Scotland *perhaps* as long as anywhere else, as we find the name—in the form of *Capull-coille*[1]—in old Gaelic songs (see Part II. *antea*, pp. 3, 27).

Eggs were sent to the Black Mount, by Mr. James Guthrie, at Lord Breadalbane's request, but none were hatched out. I am informed, however, that a bird was seen there about 1867 or 1868.

[1] This is the name handed down, and still used in Lochaber and adjoining parts of Argyleshire, districts which—with Badenoch, Atholl, Breadalbane, and Upper Braemar—have retained in greatest purity the Gaelic topography of Scotland, as we are informed by Robertson in his work before quoted.

In 1870 two males seem to have done a good deal of prospecting in Argyleshire and adjoining parts of Inverness, having been first heard of and seen in Ardgour (about 46 miles from ⊙), at Callaob, on Loch Leven, and a few days afterwards at Camus-na-gaul, opposite Fort William. They were then seen—always supposing them to be the same birds, and we cannot fix a limit to the distances to which males will wander in search of mates—in Glengarry (Inverness-shire, *q. v.*) The Rev. A. Stewart, Nether-Lochaber, considers that these birds probably entered Glengarry and Loch Ness by Kingussie, Loch Laggan, and the valley of the Spean, but I have failed to discover the localities whence they could have come by this route.

The Rev. A. Stewart (who has written many excellent articles to the '*Inverness Courier*' under the '*nom de plume*' of "Nether-Lochaber,") proposes to suggest to "Lochiel" the propriety of introducing a pair or two to the woods of Loch Arkaig side, near his residence—Achnacary Castle.

In 1875 birds arrived of their own accord at Ardkinglass, and are supposed to have come direct from Breadalbane, doubtless through Glenfyne, having forced their way up Glen Dochart, and overflowed from the internal great pressure at Taymouth (*see* general remarks under Glen Dochart further on, p. 109). At Ardkinglass there are about 550 acres of suitable woods, and the birds are preserved and increasing. As many as twelve were seen one day this year (1878). A hybrid, killed in October 1878 at Ardkinglass, was forwarded to Mr. Robert Small, Edinburgh, for preservation. It was killed by the keeper, and is now in the possession of Mr. Brodie, Edinburgh, who leased Ardkinglass shootings. No difference is observable here in the numbers of black game, which are fairly abundant. This is an interesting example of the great pressure at a centre forcing birds over a bare and unfavourable country (see general remarks *ut sup.*, and under Kincardineshire, *infra*, p. 95).

It is considered by a correspondent who is personally acquainted with the lie of the valleys between Crianlarich, Tyndrum, and Loch Fyne, that the probable route taken by the birds was from Tyndrum, skirting the northern base of Ben Laoigh, where there are some fragments of old (indigenous ?) Scotch fir, and then, on rounding the western base, a point is reached which looks down on Loch Fyne, and also through the Quurn—or Cairn—Pass into Glenfalloch. But the distance from the straggling old Scotch firs on Ben Laoigh to Loch Fyne is only between three and four miles, and this latter would be the country which would *first* present itself to the gaze of the pioneer birds. I am also informed by another correspondent—W. Colquhoun, Esq.—that his brother thinks that there are no woods which would lead them from the old fir wood near Crianlarich by the Bhalloch into Glenfalloch, and thence by the latter and across into Glenfyne, so I think the route already indicated will prove to be the most likely. By this route also the bird reported at the head of Loch Lomond (*v.* Dumbarton, 1878) would probably arrive.

About 1876-77 they made their appearance at Inveraray. Two or three cocks and one hen were the first birds to arrive. The hen killed herself on the deer-park fence in April 1877. No more hens have made their appearance. The most seen in one day were four cocks together, but there are a good few hybrids; one of these latter was killed lately by flying against the telegraph wires. "Last season I got eggs from Arran, and put them into grey hens' nests, and they hatched, but I am sorry to say I have never seen any of the birds. If all is well I shall try the same next year. The woods here are quite adapted to the bird's habits; and are four miles long by an average width of half a mile. The Duke is very anxious to get a few more. It is generally supposed that they came here from Taymouth, but whether from Loch Awe side or by Glen Fyne will be difficult to say. Ardkinglass is only five miles as the crow flies from here. They could have crossed the head of Loch Fyne, which is only about half a mile wide at Ardkinglass, and thus they would have trees all the way down to Inveraray." For the above useful notes I am indebted to Mr. J. Thompson, head keeper, Inveraray. I think there can be little doubt about the route by which they have arrived —viz., by the Ardkinglass and Glen Fyne route from Glen Dochart.

This overflow from Glen Dochart is made all the more worthy of attention by the apparent preponderance of males. The hybrid killed by flying against the telegraph wires—above-mentioned—is a male, indicating the more unusual, in my opinion, *male* parentage of *Tetrao urugallus* male, and *Tetrao tetrix* female. This goes far to prove the extreme wandering propensities of males, and to account for even more curious records than those at Fort-William in 1807, and at Bo'ness (*v.* Part II. p. 26), not to speak of the prospecting party of males which travelled over so much ground in Ardgour and Glengarry.

INVERNESS.

In Hartings' '*Sale Catalogue*' of his collection of eggs (Stevens', 6th June 1872, p. 20) occurs the entry "Capercaillie —*Tetrao urugallus*—two; Guisachan, Inverness, May 1868. E. Hargitt." They were introduced about that time, young birds having been presented to Sir Dudley Marjoribanks by Mr. Mercer of Gorthy, near Methven, Perth; and the year previous, eggs were sent from Perthshire to Guisachan. I learn from one correspondent that all the birds at Guisachan have died off.

In 1873 birds were introduced at Invereshie, Inverness-shire, but were never fairly established. A cock was shot in 1875, and at present—1878—only one hen is to be seen. There is abundance of suitable cover, extending four miles in one direction, and half a mile broad; and the present keeper has instructions to introduce them again next year—1879—by birds.

Former occurrence of the species in this county is mentioned under previous heads (*antea*, pp. 16, 20, 26).

ABERDEEN.

For an account of an unsuccessful attempt at introduction at Mar Lodge, see *antea*, pp. 37-42. The Capercaillie is not spoken of in MacGillivray's '*List of the Birds of Deeside and Braemar*,' (1853.)

For notice of its former occurrence on the "Brea of Marr," see under previous heading, p. 18.

In 1873 eggs were hatched out at Inverernan, Strathdon; but some of the birds eventually found their destination in the Museum of the Marischal College of Aberdeen, having been sent to Mr. Robb, the curator, for preservation.

Any other statistics I hold from this county are strongly negative. I have elsewhere pointed out what I believe to be the most natural avenue into Aberdeenshire, but it is also possible that another route may be chosen by the birds, viz., by Glen Tilt from Blair Athole; but at this point there are some 10 miles of unsuitable country intervening, and I rather incline to the route by Kincardineshire (*q. v.*)

CHAPTER XIV.

EXTENSION IN KINCARDINESHIRE.

At present the only record I have of its occurrence in the county is that of a hybrid, which was shot at Fetteresso on 26th October 1872 ('*Scot. Nat.*' 1873-74, vol. ii. p. 57; and Mr. George Sim, *in lit.*) It is not easy to decide the source of this hybrid, or whence came the Capercaillie which produced it. As far as I can learn at present, Stracathro is the locality nearest to Kincardineshire where Capercaillies are found (see Forfarshire, *antea*, p. 73).

Since the above remarks under Kincardine and Aberdeen were penned, I am glad to be able to record that birds have occurred on Deeside, at the very spot where I anticipated they would first make their appearance.

In August 1878 three Capercaillies—one old hen and two young—male and female—were shot in the pine woods on Scoltie Hill, near Banchory, and not far from the junction of the river Feugh—half of the waters of which come down the wooded strath of Glen Dye—with the river Dee. Mr. Charles Danford, who sends me the information, adds: "This is the first time I have heard of them in this district;" and I have a negative return from the estates of Durris, lower down the Dee, in which my correspondent—Mr. C. M'Hardy, forester at Durris—states: "We have none of them in the north that I am aware of. . . . I am aware of their having been tried in

Strath Don, where I lived previous to coming here, but the attempt failed."

I am particular in my record of this occurrence, because I am inclined to look upon it as the commencement of a new era in the history of the Capercaillie in Scotland, their future line of advance, probably rushing up the great wooded valley of the Dee as far as the pine woods reach—*i.e.* almost to the base of the Ben Muich-Dhu, and thence—if we pursue them farther—surmounting the confines of the valley, *northward* into Strath Don and the north of Aberdeen, into Banff and Inverness. I could even now point out the probable avenues of advance *out of* Aberdeenshire, but it would be, perhaps, premature to do so until we see with what success they populate the Dee valley.

I am particular in recording this occurrence for another reason also. I have failed to obtain evidence of the presence of the bird in any other localities in the county besides those mentioned, and I have received several negative returns, along with the general statement in most cases, that it appears to be almost unknown in the county. This is curious, seeing that, since the year 1865, they have been present at Stracathro, just on the borders of Forfar and Kincardine. Thus, a distance of some 18 or 20 miles as the crow flies, has been bridged over (although at intermediate localities there is abundance of forest growth to act as *stepping-stones*), and a watershed has been crossed. This finds a parallel in the overflow from Glen Dochart into Argyleshire (*q. v.*), but it is difficult to say which instance is the more remarkable.

In August or September 1878, one bird was shot near Inchmarlo House, the only one ever seen there. The woods are continuous and extensive between Banchory and Inchmarlo, and the whole parish of Banchory Ternan is almost entirely covered with pine woods, in every way suitable to the successful restoration of the birds to Deeside.

In 1878. The first has also been killed upon Fasque estate, the property of Sir Thomas Gladstone, Bart. This forms a *forest stepping-stone* to the advance into Deeside.

In 1878 another has been shot at Fetteresso, and it has also been lately found on Inglismaldie. The dilatory advance in this direction is most remarkable when it is remembered that the country is well wooded on both sides of the watershed.

CHAPTER XV.

EXTENSION IN ROSS, ELGIN, AND COUNTIES OF THE MORAY FIRTH.

THERE is a vague idea that they existed about 100 years ago in Ross-shire, but I can obtain no satisfactory statistics.

I am not aware that any attempts at introduction have been made in Ross-shire, but Captain Dunbar Brander, of Pitgaveny, has informed the Rev. George Gordon that Lord Fyffe had obtained eggs of the Capercaillie for Lochnabo—a wood between Elgin and Fochabers—this last summer (1878), but the results he had not learned.

CHAPTER XVI.

EXTENSION IN SUTHERLANDSHIRE.

For the following account of the attempt at introduction in Sutherland I am indebted to Mr. Thomas Mackenzie, sheriff-substitute, Sutherland. "It may interest you to know the result of an attempt made in 1870 by Mr. Chirnside of Skibo to introduce Capercailzies into Sutherland. A setting of ten eggs was obtained, I think, from Perthshire, and these were all hatched out at Skibo. Five of the young birds were handed over to the care of the game-keeper of Mr. Gilchrist, of Ospisdale, the adjoining property, but all of these died within three weeks. Of the five left at Skibo, three arrived at maturity, when, unfortunately, a weasel attacked and killed one of them, and during the hubbub and confusion occasioned by this, the cock bird would appear to have flown into the kennel of dogs, where it was destroyed, for the feathers were afterwards found there. The remaining bird, a hen, frequented the woods about Skibo Castle for two years after this, and was last seen in the garden in July 1872, after which it finally disappeared. What its ultimate fate was I cannot learn with certainty, but there are no Capercailzies at present on either of the properties of Skibo or Ospisdale."

Mr. Mackenzie further remarks—"For an experiment on so small a scale the result was not unpromising, and the survival of a single bird for upwards of two years, would point to the general suitableness of the locality, and to the prospect of a renewed attempt being attended with better success. The great mortality which occurred immediately after hatching was owing, I fear, to the eggs having been set under such an ignorant foster mother as the common barn-door hen, and I would suggest that, as a remedy, the experiment might be tried of placing three or four eggs in the nest of the wild grey hen, to which species the Capercailzie is so closely allied, and leaving her to bring up the young birds. From what I can learn, there is no antipathy between the two species." Mr. Mackenzie further informs me that there are some 1800 acres of suitable wood, principally fir, upon the Skibo and Ospisdale estates.

Of its former occurrence in this county I have already spoken (*vide* pp. 19, 29).

CHAPTER XVII.

EXTENSION IN SOUTH OF SCOTLAND.

AYRSHIRE.

AN attempt to introduce the Capercaillie into Ayrshire was made in 1841 or 1842, both by birds and eggs, at the aviary at Glenapp, belonging to Lord Orkney. All the eggs were hatched out, and in 1843, in September, nine healthy birds were seen there by the author of '*The Birds of Ireland*' (*q. v.* vol. ii. p. 33). But by the year 1848 they again became extinct, in which year the last of these introduced birds was found dead in the open on the farm of Downan. The birds and eggs brought to Glenapp came from Taymouth. The plantations on Glenapp were at that time too small and too scattered to keep them, and the birds either wandered and were shot down, or died, succumbing to the adverse conditions around them. It is thought that, now the plantations have increased and grown to a suitable size, if another introduction were attempted it would prove more successful.

Lord Ailsa's gamekeeper at Culzean, obtained eggs from Arran, and got some of them hatched out, but none of the birds came to maturity.

A bird was reported to have been killed on the 12th August 1877, at Old Cumnock, by Captain R. M. Campbell, but I have failed to obtain proper authentication of the fact; indeed, he himself cannot corroborate it.

GALLOWAY AND WIGTON.

In 1869, at Newton Stewart, a female bird was shot (*vide* '*Scot. Nat.*,' vol. i. p. 44). Probably a strayed bird from the Glenapp introduction in Ayrshire, or the Sanquhar introduction in Dumfries. (See also R. Gray—"On the present distribution of the Capercaillie in Scotland;" '*Proc. Nat. Hist. Soc.*,' Glasgow, vol. ii. p. 10; read December 28, 1869.)

LANARK.

In 1868 a fine male was shot by Henry Lees, Esq., at Auchengray, near Airdrie. Probably wandered from some locality where attempts at introduction had been made. (*Vide* R. Gray, *op. cit.*)

KIRCUDBRIGHT.

1869. In November 1869 a female was killed at Auchencairn. Probably a strayed bird from Glenapp or Sanquhar ('*Scot. Nat.*,' vol. i. p. 44).

DUMFRIES.

Mr. Lindsay, the Duke of Buccleuch's gamekeeper at Sanquhar, tried to introduce them, but failed. (*Auct.* R. Gray, *op. cit.*)

CHAPTER XVIII.

ARRAN.

We now come to Arran, the centre of a separate introduction. Its isolated position, the fact of its being formerly frequented by the species prior to its extinction in Scotland, and of the successful re-introduction and self-stocking, gives it a right to be considered apart from the other centres of introduction on the mainland.

It is doubtful, I think, if any localities on the mainland can come to be populated naturally by Arran birds, *i.e.*, from a natural outward pressure of population, so that, though a centre, it cannot be expected to exert an impulse of similar extent to what Taymouth has done.

The Capercaillie formerly abounded in Arran, as we are informed by the Rev. J. Headrick (' *View of Arran*,' 1807), but had apparently become extinct at the time he wrote. It was re-introduced in 1843, and for some time the birds were kept in an enclosure near Brodick Castle. It became common, and spread to the limits of the fir plantations, while an odd bird occasionally wandered as far as Lamlash, by 1870. It became still more abundant in 1872 (R. Gray, ' *The Birds of Arran*,' Glasgow, 1872), "when thirty to forty birds might be seen in a day's walk, and twenty nests be found in a season."

Mr. Croll was head keeper at the time of their introduction at Brodick, and had charge of them. Mr. George Croll, his son, has supplied me with the following account of the introduction:—"The first introduction came from Taymouth Castle in 1843. The birds numbered six hens and one cock. The second introduction came from Sweden to London in 1846, and I went there (London) to receive them, and brought them to Arran. The number of birds, eight hens and two cocks. The approximate estimate of birds in after years would be, in 1855, about 40, and in 1865, the year I left the island, about 65 or 70. I consider that the birds cannot increase to more than 80 or 90 birds, owing to the limited extent of wood and planting. Black game have decreased on the island of late years, not on account of the numbers of the Capercaillies, but solely through the want of cropping on the moor-edges, which have been turned into pasture. There has been little or no planting of Scotch fir, larch, or spruce, since the year 1830."

The birds are confined to the neighbourhood of Brodick, Arran, where there are about 600 acres of fir wood; only stragglers having been found in other parts of the island.

PART V.

GENERAL REMARKS ON THE DISTRIBUTION AND INCREASE OF THE SPECIES;

AND

ON THE DECREASE OF BLACK GAME.

CHAPTER XIX.

LAWS OF EXTENSION OF RANGE.

THE general distribution of the Capercaillie throughout the world is concisely given by Mr. H. E. Dresser in his great work '*The Birds of Europe*,' part xxi.[1] He says:—"This magnificent grouse, the largest of its family, is found throughout Northern Europe, and is even met with in the forests on the mountain ranges in the southern or southern-central portions of the Western Palæarctic region." Perhaps the most southerly locality recorded in Europe is Acarnania in Greece (*op. cit.*)[2] To the east it extends far into Asia, and has been found to occur as far as the valley of the Irkut; but in the extreme east of Siberia it becomes partially, if not wholly, replaced by a smaller species—*Tetrao urugalloides* of Middendorf. Beyond this, in the present connection, it is unnecessary to enter into detail as regards its distribution outside the limits of Great Britain.

Within the area of its present range in Scotland, suitable woods—*e.g.* woods of spruce, Scotch fir, or larch, or of these combined—of 100 acres, or even less in extent, and upwards, are usually inhabited by the species; the smaller woods hold-

[1] See also Lloyd's '*Game Birds and Wild Fowl of Sweden*,' 1866, p. 2.

[2] Pennant, '*Arctic Zoology*,' 1792, vol. i. p. 365, seems to have traced it as far south as the Archipelago, in the islands of Crete and Milo. Hasselguist is given as the authority for the bird shot in a palm tree in Milo, and Belon for Crete ('*Penny Cyclopædia*,' vol. vi. p. 260).

ing only a few pairs, sometimes permanently, but often only for a few years consecutively;[1] and the larger woods and continuous forests holding more in proportion according to their areas, as many as 36, 25, 23, etc., having been shot in one day by shooting-parties upon certain estates in Perthshire.

But in districts inhabited by them, where forests, once extensive, have been reduced in size, or entirely cut down and replaced by sheep-grazing on the cleared ground, a reduction in numbers or entire disappearance naturally takes place.

Change of residence or local migrations, such as are mentioned by several writers—*e.g.* Lloyd, Collett, etc.—I have myself observed also in Scotland, and many of my correspondents draw my attention to the fact in their letters. The birds often entirely quit one part of a forest or extensive wood for another, and this migration usually takes place from an older to a younger growth—*i.e.*, to a growth which is of that age which is suitable to their requirements of feeding and shelter combined. There is also a decided preference shown by the birds for forest slopes facing the south, or at all events for those lying well to the sun, as I think can be traced by the general distribution of the birds at the present time. I find that in various localities they appear to thrive better and to increase more rapidly where they inhabit situations having a southerly exposure; other amenities, such as the necessary amount of cover and absence of disturbing influences, being of course taken into consideration. The above-

[1] Thus, at Aberuthven wood, on the south side of the Earn valley, or right bank of the river, a pair of Capercaillies bred regularly for four or five years, but, along with their young, always left after the latter could fly. Since about 1869 they have not returned to this locality—a wood of about 40-50 acres in extent. I could give many other instances of their breeding sporadically in this way, and also of their breeding regularly, but always leaving with their young for larger covers afterwards. This is regularly the case at Ochtertyre, Crieff district, where, I am informed by Sir Patrick Keith Murray, Bart., "nests are found all over the woods and copses, miles from the winter resort of the birds."

mentioned partial migration or change of residence may, in many cases, be induced by this preference. (See also, under "Restoration," the account by Sir Robert Menzies, Bart., of the preference shown to the woods of Dull, near Taymouth.)[1]

In certain districts, as, for example, in Glendochart, in the west of Perthshire, Capercaillies are known to frequent coppices of hardwood (birch and oak), and even to occur regularly in autumn at some distance away from wood of any kind, being often shot by sportsmen as they rise out of long heather on the hill-sides.[2]

In the winter season, however, most of these latter repair to more suitable shelter.

Their favourite haunts are spruce, Scotch fir, or larch forests, and their occurrence in hardwood coverts is comparatively rare, except in the breeding season, or in the case of birds resting during a tour of inspection. The departure from their usual habitat, on comparatively rare occasions, may be accounted for by a natural impulse urging them to extend their range, notwithstanding unusual difficulties and unsuitable ground, combined with an unusually strong pressure outward from the nearest centre of population. An unusually strong pressure of this kind probably takes place outward from Taymouth; and Glendochart being the direct

[1] Correspondents inform me that the young birds do not like the sun, and that in the heat of summer they often shelter themselves under overhanging banks, apparently to escape from it. In the heat of summer even the old birds sit more upon the ground than earlier or later in the season. In Glenalmond, in Perthshire, the bank of the river facing the north is preferred, but that is simply because the distribution of the pine woods favours them. On the south side of Loch Rannoch it has been found rather difficult successfully to rear and keep them. The young birds hatched out at Cromer Hall (*vide* under "Restoration"), supposed to have died from exposure to a scorching sun ('*Penny Cyclopædia*,' vol. vi. p. 260), may, however, have perished from an insufficient supply of their natural food.

[2] In Norway, Sweden, and Russia, the Capercaillie also occurs occasionally in hardwood coverts—oak, beech, birch, etc.—but is not usually stationary in such localities; but, as remarked by several Scandinavian naturalists, seems fond of patches of such growth in the midst of pine forests.

outlet towards the west, the wave of advance has been forced up the valley until it has reached beyond the limit of spruce, Scotch fir, and larch. The inevitable law exists that there is a distinct limit to the population of a species in any one locality, regulated by the size, capabilities, and amenities of the area, and when this limit is reached and exceeded the surplus population is forced to seek new ground. Further, as the internal pressure continues and increases, the greater difficulties will be faced and surmounted by the pioneers, in order to fulfil the destiny of the species, and the amount of success achieved will depend upon the hardihood and "fitness" of the species in the struggle for existence.

The easiest and most natural courses of advance follow the valleys, up or down stream, or stretch along the wooded foothills and slopes, preferring, as already pointed out, the sunny exposures. This is clearly illustrated by the fact, that Capercaillies in a comparatively short space of time reach and populate more remote localities in the directions whence the great valleys lead them, than they do in those directions where mountainous or treeless country intervenes.[1]

The beau-ideal of a safe harbourage and permanent home for the Capercaillie would then appear to be as follows:—A forest of spruce, Scotch fir, and larch, mixed, of, say 700 to 1000 acres or more in extent, having a southern exposure, or lying well to the sun, of which forest certain portions—say a

[1] The comparative rate of advance by valleys and over mountains is illustrated—to choose amongst a great mass of similar statistics—by the extreme limit reached by the Tay valley route in 1877 (*vide* Map), and the distance reached at Loch Kennord in the same year; the former—Milngavie near Glasgow—some 45 miles in a direct line from Taymouth, and the latter only 7½. Or, to select another example—the limit reached by the Tay valley route to Dunkeld, and thence by Strathmore into Forfar in 1865 (Brechin district—over 50 miles in a direct line from Taymouth), as compared with that reached in Glenshee in the same year (Dalnaglar, only about half that distance in a straight line from Taymouth). I am inclined to think, judging from a large mass of data, that the birds rarely, or only under exceptionally strong pressure, cross mountain tracts, *unless they can view forest-covered country beyond.*

fourth or a fifth—consists of old timber, and the remainder of a succession of growths in regular rotation; forested in fact like a German forest, thus offering abundance of food, shelter, and quiet. And further, there ought to be, so to speak, convenient natural avenues or wooded continuations—'*forest stepping-stones*'—to other districts, either in valleys, or along the hill-slopes, to act as safety-valves for the escape of surplus population.

The hen Capercaillies appear to be, in most cases, the pioneers, which lead to the extension of the range of the species; and it is natural that they should act as the pioneers, as they are more numerous than the cocks, and increase in numbers more rapidly, as is the case with most polygamous species.[1] The birds, from some point of vantage on the outskirts of their residence, view a large pine wood, even at some miles distance, and make direct for it.

It is suggested and believed by several correspondents that the pioneers are *entirely* composed of young birds driven away from the *lecking* ground, and haunts of their native coverts. This is no doubt in great measure the case, and is only part of the mode in which the natural law is put in force, but I must certainly uphold that attraction does take place by the most likely coverts and woods, as all our statistics indeed go to

[1] According to Lloyd—quoting the writings of others on the birds of Sweden—it would appear that the males are by many considered the most abundant, giving rise to the extraordinary migrations of the male birds from time to time, '*en masse*,' in the north of Europe; and we also are told of the wandering habits of the males, which are said to '*forflyga sig*,' *i.e.*, "to fly it knows not whither"—in the same way as I find the hens do in this country—and being shot in strange out-of-the-way localities. The author of '*Tidskrift for Jagare*' is quoted by Lloyd, as stating that from experience he "has found that both capercali and blackcock broods contain more males than females." Certainly this flocking together and wandering propensity of the males in Scandinavia is singular. I have utterly failed to obtain any statistics confirmatory of this superabundance of males in Scotland; indeed, our experience is quite opposed to that of continental naturalists, unless, indeed, the killing of 17 males in one day at Keillor wood, near Methven, can be said to point to a superabundance of males.

prove; and this, I think, will become patent to any one who studies their distribution. If, on the one hand, they are *forced* to leave by the older and stronger birds, still they will leave in those directions which are most likely to meet the requirements of the species, and, as I have already pointed out, I believe that the birds have great power of vision, and use this in an appreciable degree when on a pioneering journey.

Birds, *and especially females*, are thus often shot or seen in localities totally unsuited to their habits—no doubt resting, as I have already indicated, during their tour of inspection. Amongst such localities may be instanced a bare moor or open common,[1] a patch of wood, of an acre or two in extent, in the middle of a bare mountain glen,[2] or even in the crowded thoroughfares of a large town.[3] There is evidence in some cases of these pioneers having been assisted in their travels by long-continuing gales. Thus, about the first bird shot in Fife—in 1863—at Rankeillor, near Cupar, arrived towards the end of a gale which had been blowing for some days from the north; and several other instances could be cited.

The females precede the males by from one to two years, and establishment of the species takes place very shortly after the arrival of the males, and from two to four years after the first appearance of the females—*i.e.*, where establishment *does* follow (*vide* Tables given below).

In the comparatively few instances in which males are first observed, it may be inferred, in most cases, that the females had arrived from one to two years previously, and had escaped observation, or that the males had wandered during their search for their pioneers. In certain districts, where there is only a limited population at the centre, a

[1] Stenhousemuir, Stirlingshire, for example—a bare grass common, with a few scattered whin bushes—used to hold markets upon.

[2] As Glen Queich, and many others in Perthshire and other counties.

[3] In Edinburgh (*vide* under Midlothian, *antea*, p. 89).

natural impulse causes the males to go in search of the hens, and it is probably while so doing that they are said to "*för-flyga sig*," or "fly they know not whither," in Sweden. Witness the occurrences of a male bird at Auchengray, in Lanarkshire, and (whether prior to extinction or not) of the male birds shot at Fort William in 1807, and at Bo'ness in 1811. The following table shows some of the above facts. The statistics are selected from a very large number in my possession, all going more or less to prove what I have above stated.

TABLE SHOWING ORDER OF ARRIVAL AND ESTABLISHMENT OF THE SPECIES.

Name of Localities.	Date of arrival of Females.	Date of arrival of Males.	Established.	Time between first and last dates
Cardross, Perth . .	1868	1870	1871	3 years
Chasefield, Stirling .	1875	1877	1877	2 ,,
Boquhan, do. .	1871	Bred	1875	4 ,,
Lanrick, Perth . .	1856	1856[1]	1859–60	3-4 ,,
Moncrieffe, do. . .	1861	?	1864	3 ,,
Kinnordy, Forfar . .	?	1863	1863	?
Fotheringham, Forfar	1862	1862	1862	Same year
Torwood, Stirling .	? Birds arrived 1867	...	1870	3 years
Auchengray, Lanark.	?	1868[2]	Not established	
Rossie Priory, Stirling	1867-68 }	No males[3]	Not established	
Rossie Priory, do. .	1877-78 }			

[1] The remains of a male were also found in 1856 at Lanrick, or Gartencaber (adjoining). Its death, from unknown causes, very possibly delayed the establishment a year or two. Otherwise, judging from analogous cases, the establishment would probably have taken place in 1857. A separate informant gives 1853 as the date of their first appearance at Lanrick; but judging from parallel records at contiguous localities, I think this may be too early stated, though, of course, quite within the bounds of possibility.

[2] This male, killed in 1868 at Auchengray, was doubtless a bird wandering in search of a mate from some one of the localities south of the Clyde, where several futile attempts at separate introductions were made.

[3] A very sufficient reason here for the non-appearance of males is the insufficiency of cover. Although the females twice attempted to extend the range of the species in this direction, they were not followed up by males; or it may possibly have been because the hens were in both instances shot soon after their arrival.

The Capercaillie, then, has populated the woods and forests of part of Scotland, principally by its own exertions, since the great restoration at Taymouth; but there are certain minor centres of introduction which have undoubtedly added *some* impulse to their advance, though, perhaps, not to any extent compared with the impulse from the great centre. Thus, we have the Tulliallan introductions in 1856 and 1864, and the Arran introduction, and that at Lathirsk, in Fife, besides others more or less successful or unsuccessful. The Arran introduction, perhaps, has done least to extend the bird's range, its isolated position naturally placing a bar to their spreading, although the introduction in itself has proved eminently successful, so far as the limited area of wood permitted.

CHAPTER XX.

A FEW REMARKS ON HYBRIDISM.

ON first arrival and establishment of the Capercaillie at a new locality in an entirely new district, where black game are abundant or fairly plentiful, cases of hybridism are not unusual. If females arrive first—as we have shown they do—in a district populated by black game, the absence of their natural partners will induce coition with black game, and will result in hybrids. If the male Capercaillies are long of arriving, this hybridism will increase in frequency. But when their own lords at last make their advent, the hens undoubtedly, for the most part, return to their allegiance, and hybridism becomes rarer, though it may not altogether die out. But, as we have already seen, the males *usually* do arrive very shortly after the females—a scarcity of female birds at the centre, caused by the overflow, doubtless inducing the males to follow. Only in unusual cases of isolation, or unusual distance from the centre, do the males fail to find out the hens. Thus it is only rarely that hybridism attains to serious proportions. As the exceptions often prove the rule, I will instance a few statistics chosen from amongst a number of others.

At Logiealmond, Perthshire, two hybrids were shot in 1852, the first obtained there. Also a hybrid was shot at Alyth, in the east of the same county, in 1857. At Tulliallan the first bird ever seen there was a hybrid in 1854. These three localities are situated respectively about 17½ miles as the crow flies, 28 miles, and 38 miles, from the centre of restoration—Taymouth; and these were amongst the birds which had reached to unusually great distances at these dates. I cannot be far wrong in supposing that female Capercaillies first reached these unusually distant localities, it may have been assisted by gales of wind (as in the case of the first recorded Fifeshire bird (p. 76), and that male Capercaillies failed to follow them. Landing in a country inhabited by black game, hybridism resulted.

Again, at Tulliallan, in 1856 (two years later), out of a setting of Capercaillies' eggs three birds were reared, but these turned out to be all females, which "bred freely with black game, and hybrids were common in 1857. In 1864 more eggs were hatched out, some of which were males, and since then Capercaillies have increased rapidly, and now—1877—there will be from 200 to 300 birds on the estate. Hybrids are still to be met with, but not so numerously as before the Capercaillies became plentiful." For information, so much to the point, I am indebted to Mr. Millar, head gamekeeper at Tulliallan, who has been there since their earliest appearance. The eggs hatched out in 1864 were obtained from Freelands, near Perth, by Lord Balfour of Burleigh.

Sabanaeff, in his account of the '*Avi Fauna of the Ural*,' says, under '*Tetrao urogalloides* (*T. Medius*)':—"Taking this as a hybrid, it is easily explained, as a great number of the male *T. urogallus* are killed in spring, and therefore there exists a great predominance of females" (see translation of his paper in '*Proc. Nat. Hist. Soc.*' Glasgow, 1877, p. 304). I am not sure, however, that this is entirely the reason of a predominance of females. Farther north, at Ust-Zilma, on the Petchora, Seebohm and I found that the natives only shot the *hen* Capercaillies, as the males were not considered fit for food.

As to the occurrence of females in whole or partial male dress in Scotland, I know of the following:—One in the possession of Lord Balfour of Burleigh, at Kennet, in Clackmannan, shot on the 2d November 1862, near Dunkeld, at Mr. Hugh Bruce's residence. This specimen "shows its tertiaries and scapulars tipped with white, and so far slightly resembles the plumage of the adult female."[1] This was the first example of the kind recorded in Scotland, but another has since been obtained, also shot at Dunkeld, which is now in the Museum of Science and Art. It was obtained at Dunkeld in 1866, and purchased from Mr. Robert Small, naturalist, Edinburgh, for the Museum.[2] Mr. J. H. Gurney *jun.*, informs me that he has in his possession a dwarf Capercaillie in nearly full plumage, bought in Leadenhall Market. This is probably of Scandinavian origin, however. Mr. Gurney adds:—"There is a similar dwarf in the British Museum, but larger than mine. It may, for aught I know, be a female in male plumage. It is labelled, 'Hybrid grouse, *var.*,' but I do not believe in its being a hybrid." In our collection at Dunipace is a very diminutive Capercaillie in male plumage, shot at Dunira, and given to me by Mr. J. Hamilton Buchanan, which I believe to be a female in male plumage, but unfortunately the specimen is not sexed.[3]

[1] *Vide* '*Proc. Ryl. Phyl. Soc.*,' Edinr., vol. viii. 1862-63, pp. 25-27.

[2] *Op. cit.* 1865-66, p. 408.

[3] For full particulars regarding hybridism and varieties, etc., see Collett, '*Birds of Northern Norway*,' quoted very fully by Dresser, '*Birds of Europe*,' parts 20 and 21. In the Christiania Museum is a truly wonderful series of the lovely varieties found in this species, which will always well repay the attention of visitors. I had the pleasure of examining these in 1871. For an account of an unusual hybrid between black grouse and hazel grouse, see Dresser in '*P. Z. S.*,' 1876, p. 345. There are many other notices and accounts of hybrids scattered up and down, and a general reference to ornithological works might suffice in this place. However, to those who feel curious on the subject, I may recommend an examination of Carus and Engelmann's '*Bibliotheca Historico-Naturalis*' (vol. ii. pp. 1253-54), which will supply a list of papers, British and Foreign, up to the date of 1861.

CHAPTER XXI.

ON THE INCREASE OF CAPERCAILLIES.

My circular contained several queries, specially bearing upon this part of the subject (*vide* circular, p. 54, queries 9th to 13th). From the answers received, I have put together the following statistics, as they may be useful for future comparisons.

Perhaps the districts where the Capercaillie is most abundant at present are as follows:—The Tay Valley from Taymouth to Perth, and the districts between Perth, Crieff, and Comrie, in Perthshire; Strathmore, in Forfarshire; Tulliallan, in south of Perthshire; Sauchie, in Stirlingshire; and Arran.

The greatest number I have heard of killed in one day was 36 at Ladywell plantation, Pitnacree, near Dunkeld, in 1865. This is part of the Athole estates, upon which are at least 10,000 acres of suitable woods. At Bamff, in the east of Perthshire, 25 were shot one day in 1877—area 2000 acres. At Dunkeld 23 were shot one day (*vide* '*Game-Book*,' belonging to Mr. Small of Dir-na-Nean). At Fowlis Wester 12 were shot one day in October 1859; here the woods are extensive, as they join with Abercairney, Gorthy, Tulchan, and neighbouring properties. At Methven 14 were shot one day—300 acres. In Keillor wood, Methven, as I am informed by Mr. D. Brown, he once saw 17 cocks killed in one day.

At Ochtertyre great numbers were seen in 1870 or thereabouts. At Cardross, where they only arrived in 1868 and 1870, and became established in 1871, and where there are about 300 acres of suitable wood along the south shore of the lake of Menteith, 18 to 20 may be now seen in a day; and in November 1877, 4 males and 6 females were shot one day. On Murthly 12 were shot one day, and from 20 to 25 in four days—about 3000 acres of fir and larch. Here they became scarcer for a time after a certain date, and a correspondent assigns over-shooting on a neighbouring property as the cause. Now they are increasing again. On Tulliallan it is reckoned there are "from 200 to 300 birds," in woods of an area of at least 1500 acres. On the adjoining property of West Grange I have myself seen at least 16 birds in one day, and seen 4 shot. At Easter Ogle, in Forfarshire, 20 to 24 have been shot in one day—about 2000 acres. On Sauchie, in Stirlingshire, there are estimated to be from 50 to 60 birds. At Dunsinane, near Perth, in 1868 or 1869, I saw upwards of 30 birds in one day. At Dupplin 16 were shot one day. At Taymouth, about 1862, estimates reached between 1000 and 2000 birds.[1] At Torwood, in Stirlingshire, in 1878, about 14 were driven up to the guns in one beat, and 3 were shot.

The above are only selections from a larger mass of materials, but they will, I think, fairly illustrate the districts in which the birds are most abundant. The farthest point reached towards Loch Earn is Dunira, except a stray bird or two on Loch Earn side at Ardvorlich and Dearry.

[1] According to returns to Parliament in 1873, the total acreage of the Taymouth estates in Perthshire was 234,166; of this, of course, only a comparatively small proportion is under wood.

CHAPTER XXII.

ON THE DECREASE OF BLACK GAME.

BLACK GAME have become less plentiful, and in some cases even extinct, or nearly so, in certain districts where Capercaillies have become plentiful, and there is an *apparently* striking connection between the advent of the latter and the departure of the former. This cannot be denied, and I have many statements from correspondents bearing out the truth of the assertion.[1]

From many others, however, I have also the statement that black game have not decreased in certain other districts which Capercaillies have taken possession of, and where they have become plentiful; and the latter group of correspondents are invariably of opinion that Capercaillies in no way inter-

[1] And it does seem certain, moreover, that Capercaillies and black game do not invariably agree to live peaceably in one another's company. One correspondent, who appears to have had considerable experience of their habits, and to have watched them closely, tells me that when black game and Capercaillies do happen to frequent the same ground, the latter drive away the former in the spring time. As this, however, can only refer to somewhat unusual circumstances in the localities, and as the two can hardly be said generally to frequent similar country, this fact can only be held as of local application. The same correspondent adds that they drive away black game "by driving off the old birds and killing the young;" and he states that he has seen battles between the species in '*clucking season.*' He also saw a Capercaillie and a grey hen with their broods coming in contact, when a terrible battle ensued. "The Capercaillie succeeded in driving away the grey hen, and then killed the most of her brood."

fere with black game. From other sources I gather the fact that, in many districts remote from the area inhabited by Capercaillies, black game have become much scarcer, or have almost disappeared within the last few years; indeed, the complaint is no uncommon one, although in some cases a reason for the decrease cannot with certainty be assigned. This last fact proves, at all events, that there are other "environing causes" at work, modifying and restricting the numbers of black game, and that it is unfair to heap all the blame, if indeed any of it, upon the Capercaillie: that, in fact, it would be judicious to pause before summarily condemning the panel.

To endeavour to discover and make plain to my readers some of these other causes of the decrease of black game is the object of this section.

At Dalnaglar, a property in Glenshee, Perthshire, as I am informed by the proprietor, black game were "extremely abundant fifteen years ago." Capercaillies arrived there about 1865, when black game were still plentiful, and rapidly increased in numbers in a large pine wood of some 300 acres in extent, the largest suitable cover for the species in Glenshee. For the last ten years (prior to 1878) drainage operations have been carried on more or less: previous to that time no draining had been done for ten years. No young plantations have been formed, and all the wood on the property is now nearly, or quite, ready for cutting. Black game have decreased to such an extent that at one time they became almost extinct, and the proprietor had to re-introduce them by eggs from counties south of the Firth of Forth. I myself will not be surprised to learn some day that even the Capercaillies will change their residence and remove to more eligible covers, unless young plantations be started on Dalnaglar.

We have seen that attempts to introduce Capercaillies to

the old Black Wood of Rannoch—the remains of the Caledonian Forest—which has a northerly exposure, and where, until of late years, there has been but little fresh planting, have failed. If proprietors drain their moors and trench their forests upon the rapid system now so generally practised, I doubt if they can expect to have black game as plentifully as formerly. It is well known to naturalists that black game, especially when young and tender, are particularly fond of feeding in swampy rushy moor, or moist forest land, finding there abundance of insect food suited to their tender age. It is not until they reach a certain age that black game feed persistently upon larch, birch, and Scotch fir "sprits." Take away from them this rushy ground where their favourite, nay, *necessary* food is found, and black game will leave or die out, unless artificial feeding, perhaps, in some degree may avert the calamity; which, however, I doubt.

Again, I have it from various correspondents that in certain localities in the Tay valley and elsewhere, black game have in no ways decreased, although Capercaillies have become numerous; and in several instances I have elicited the further information that at some of the said localities draining has not taken place to any extent for a number of years.

At Arden, on Loch Lomond, as I am informed by my friend Mr. James Lumsden (who for some time back has been working at the distribution of black game in Scotland), there are no Capercaillies. Black game, nevertheless, have been decidedly on the decrease, even rapidly, during the last eight or nine years, *which is the time which has elapsed* since draining operations on a considerable scale were commenced upon the estate. There has also been a curtailing of the acreage of cropped land along the moor edges, cropping having been replaced by grazing. Cropping, I hold, of course, as an artificial mode of feeding black game; still this, at least, proves another "environing cause" for their decrease. Every year almost we hear of complaints of the decrease of black game continuing from many estates in the west of Scotland and from other parts; often, it is true, from causes at present difficult to trace, but some of which I have little doubt will be found in drainage, land improvement, and consequent destruction of insect and vegetable food necessary to the wellbeing of black game. Mr. J. B. Hamilton, of Leny, has given me a strong instance of decrease of black game, owing to land improvement, from a district not inhabited by Capercaillies, which is so much to the point that I quote his communication in full: "In Dumfriesshire I have for several years shot over the Corrie estate, which belongs to my friend Mr. Jardine, of Lanrick. It is a hilly grass country, with a deal of good strong land in it. It may extend to some 10,000 acres, and at one time he rented the shooting of adjoining land there to a somewhat similar extent. He has known that district all his life, and has shot over it for probably thirty years or more. The black game have diminished very largely in numbers during that period. His impression is that he does not see one bird now for ten that he used to do. There are no Capercaillies there, or anywhere near there, though there are plenty of fir woods of all ages and sizes. He accounts for the diminution of the black game from the improvement of the land. During the last twenty years there has been a large amount of drainage, both close and open, done on the Corrie property, and a very large amount of top dressing the hill land with lime, at the rate of from six to eight tons per acre. The result of all this is the destruction of the coarser grasses which produced the seeds that were the food of the game, and the production of a finer type of herbage, which has probably doubled the rental of that property within the last twenty years.

"Improvement of land, as a general rule, may, I think, be accepted as synonymous with injury to really wild game. Thus, I have known the size and weight of hares to have fallen off greatly on land after it has been improved; and the quality, probably, of beef and mutton, and certainly of milk, and its products—butter and cheese—is greatly deteriorated on improved land, while the quantity is largely increased."

Water is in itself necessary for the health of black game, and, as Lloyd informs us, the want of it in certain seasons "visibly diminished their numbers" (*op. cit.* p. 74). On the other hand, trees are necessary to the welfare of Capercaillies. But drainage is necessary to the growth and health of forest trees, so that the interests of the Capercaillie and black game are really antagonistic in a considerable degree to one another, though it cannot, with justice, be said that the former are actually "driving out"—which is the expression generally used—the latter. Rather say the conditions favourable to the former are, in a measure, antagonistic to the latter.[1]

Indeed, I have evidence in at least one instance of a reaction, that is to say, of the black game actually increasing in a locality where Capercaillies have been on the decrease for twenty years, not simply because the Capercaillies have decreased, but because the same reasons which caused the Capercaillies to decrease, viz., the cutting of large extents of forest without a corresponding area being replanted, and little or no draining operations having been carried on—have benefited the black game, and caused them to increase again in numbers. To make perfectly sure of this fact, I was at some pains to obtain statistics from the locality in question regard-

[1] Indeed, if we come to investigate the subject further, we find red grouse decreasing, and black game increasing, under similar or parallel conditions. Thus, in Buchanan parish, Stirlingshire, we find the red grouse "superseded by the black grouse ('*New Stat. Acct. of the County*,' p. 91); also a great decrease of the former in Kirkcudbright (*op. cit.* p. 110)." "Scarcely a black cock to be seen in 1809 in Kirkgunzeon parish, Kirkcudbright, now outnumbering the grouse" (*op. cit.* p. 219); and many more instances could be given. On the other hand, in certain districts, in localities where there is a suitable provision for both species—sufficient heather for the grouse and sufficient marshy ground and grass for black game, as, for instance, in parts of the high flat moors of Ayrshire—both species thrive and multiply.

ing land improvements, etc., during the last twenty years, and to verify reports as to corresponding increase and decrease of the two species.

In this neighbourhood also—parish of Dunipace, Stirlingshire—twenty years ago, black game were common at a locality only about a mile from this house. Drainage operations were carried out to a considerable extent, and black game entirely disappeared. Capercaillies arrived and became fairly established about 1867 in the spruce and pine woods of Torwood and Denovan, which partially surround the moorland locality in question. No drainage of any consequence has taken place for very many years. Black game returned to the locality of their own accord three years ago (1875), and have since then been increasing slowly but steadily. Capercaillies at the same time are tolerably common. The nature of the land admits of both species thriving and increasing, for while the Torwood forest has been partly drained, old wood cut, and the ground replanted, the moorland haunt of the black game has almost returned to its normal state, the same as before it was drained at all.

To open draining on estates in Scotland, and to other land improvements, injuring or destroying the natural grasses and drying up the land, is attributable, I believe, the marked decrease in black game which many proprietors lament.

Mr. Robert Collett writes me as follows:—"As to the decrease of the black game, effected by the spreading of the Capercaillie, I am sure that it must be on *very rare* occasions that these birds offend each other. They have in Norway their different localities, which they prefer to others, and their own breeding-grounds, and I have very seldom heard of any fighting between them. In Norway I never heard of any corresponding decrease and increase of the named species. Although certainly both *tetrix* (black game) and *urugallus* (Capercaillie) are straggling birds, and may be found in numbers one year in a place where they are totally absent the next." Dr. Meves writes in similar terms from Sweden.

Besides, increased population and wealth, producing a steadily increasing swarm of sportsmen year by year, together with the general use of the rapidly-loaded breechloader—has this nothing to do with the decrease of black game? I have many communications from parties well able to judge, that this *is* a factor, and a not unimportant one. Combine this with the above planting, and draining, and land improvement, and say:—Is it not more extraordinary that black game are still numerous, than that they have decreased in number during—we will say—the last thirty years?

I could bring forward much more evidence of a similar nature, clearing the Capercaillies from blame, but I think it needless further to enlarge upon this part of the subject. I will merely add in the words of an American naturalist (Prof. T. Gill, in '*The Nation*,' July 19, 1877, p. 44):—"Birds are a highly specialised group, very liable to modifications, resulting from environing causes, and ever susceptible to the changes of condition that may supervene;" and record my conviction that the increase of Capercaillies is probably the very least important factor in the decrease of the black game, if, indeed, it is one at all. Further, I believe that too often too much importance is attached to casual and superficial remarks expressed by those who do not look beneath the surface of first impressions. Error or exaggeration in consequence gains credence, and becomes, in the minds of many, even as established facts.

CHAPTER XXIII.

RELATIONS BETWEEN CAPERCAILLIES AND PHEASANTS.

Not only are Capercaillies accused of driving away black game, but in some localities (though not generally, as far as I can learn) they are said to drive away the pheasants from the feeding boxes and themselves feed upon the Indian corn. This is induced, no doubt, simply by the force of constant example set before them, and it only proves how easily our big friend can adapt himself to circumstances like a good colonist, as he undoubtedly is. Moreover, the Capercaillie, exercising his gifts still further, sometimes takes possession of a comfortably formed pheasant's nest—according to several correspondents—as, they say, three or four pheasant's eggs are found at times in nests occupied by Capercaillies. But I should doubt if in most cases such occupation by the Capercaillie is premeditated, and I would not be surprised to learn that some day or other some observant naturalist will report the fact, that the hen pheasant is really the aggressor, and second occupier of the nest. Supposing, however, that the Capercaillie really is the aggressor, how often does it happen, and what amount of harm is done? A Capercaillie hen, intent on preparing a summer home, we will suppose, wanders about in search of a suitable position, and stumbles upon a hollow in every way suited to her requirements. A hen pheasant having covered over her two or three eggs with leaves, had perhaps just left it shortly before. The Capercaillie may give it an additional scrape, or she may not, but, at all events, she settles upon it and lays her eggs. Perhaps the hen pheasant deserts, perhaps they sit amicably on the eggs by turns. In either case the eggs are kept warm, unless indeed the Capercaillie ejects them. Perhaps a gamekeeper comes by that way. He sees the pheasant's eggs, and lifts them, and they are hatched out under a hen, but he is pretty sure to blame the Capercaillie for usurping the pheasant's nest, when perhaps really the pheasant is the bird which ought to bear the judgment; in other words, the Capercaillie gets more blame than she really deserves. The number of pheasant's eggs lost in this way cannot surely be very great. When partridges and pheasants' eggs are found in the same nest—as they often are—tell me, which is the aggressor?

Other correspondents, but not many, believe that Capercaillies drive away the old pheasants and kill the young in the same way as described in the former section with regard to black game. If, as I strongly suspect, the hen pheasant is the first aggressor—laying in the hen Capercaillie's nest, and claiming a part of the maternal duties—it is perhaps natural that the stronger bird should assert her rights, or suspect the pheasant's motives, and retaliate by driving her off, and even killing the young pheasants. But, for reasons above stated, I think this can only be in exceptional cases, and only in self-defence.

There can be no doubt that the hen birds of the two species do fight at the feeding-places intended for the pheasants, and that sometimes a hen pheasant is killed or maimed in the battle; but this will find its own cure in some other way than by extirpating the Capercaillies.

PART VI.

ON DAMAGE TO FORESTS AND CROPS; AND CONCLUSION.

"Who shall decide when doctors disagree,
And soundest casuists doubt?"

NOTE.—This Part is far from complete, but I have found great trouble in getting statistics, and crops, etc., for dissection during the summer months. Perhaps the only way to elicit further data will be to publish it as it at present stands.

CHAPTER XXIV.

DAMAGE TO FORESTS.

Extensive damage done by Capercaillies to woods and forests is another sad thorn in many proprietors' sides. To hear some speak of the awful destruction going on might almost lead one to suppose that in bygone days the evil deeds of the species brought its own punishment; that, in fact, the Capercaillie exterminated the forest growth over large areas of Scotland, and that this extermination exterminated them in turn. In many parts of the area at present inhabited by the birds, on this account alone, they are shot down upon all occasions, in the endeavour to reduce their numbers; or, at all events, no encouragement is given to them to increase. Twenty years ago, when Capercaillies appeared on an estate, the greatest care was taken to foster them. Now, it is usually different; and, where at all plentiful, they are shot, both sexes indiscriminately; and in some places, as we have seen in a former section, very heavy bags are often made. I have, however, sufficient faith in the robust character of the species to prevent my becoming anxious lest a second extermination should ever take place, if fair means only be used to keep its numbers in check.

I have in this connection endeavoured to collect independent testimony from the best sources as to the nature and amount of the damage done. Mr. Malcolm Dunn (who has specially studied, and who read a paper upon the subject before the Botanical Society of Edinburgh (see '*Zool.*' 1875, p. 4338), writes to me as follows:—"In reference to the damage done to plantations by both Capercaillie and black game, the proof is too strong to admit of the slightest doubt. Where either exist in large numbers, in, or in the neighbourhood of, young plantations of larch and Scots fir, they do a vast amount of injury to the young trees, by eating the young buds, leaves, and shoots of the trees; and, in the case of the Scots fir especially, at a season of the year when there are no insects of any kind upon them—I here refer to beetles, caterpillars, or Aphides—coccus or scale insects may be upon the trees, but *not on the wood or leaves eaten by the Capercaillie.* This is the opinion of all my correspondents who have paid any attention to the matter, corroborated by my own experience. Of the very many 'crops' of the Capercaillie which I have dissected and carefully examined, none contained any traces of the bird having fed upon insects. In the *winter* the crops are usually entirely filled with the *leaves, buds,* and *young shoots* of the *Scots fir*. The contents of one 'crop' of a male bird, which I examined in November 1873, were as follows:—203 points of shoots of Scots fir, with the leading buds entire, some of the shoots being fully 3 inches long; 11 pieces of young wood, 1½ to 2½ inches long, having leaves attached, but no terminal buds; and 52 buds—making in all 266 *shoots* and *buds,* besides a *large handful* of *single leaves* of the Scots fir, which the bird had devoured at one meal. The whole were quite fresh and green, were to all appearance selected from a very healthy tree, and showed no trace whatever of ever having been attacked by the pine beetle (*Hylurgus piniperda*) or any other insect; and MOST CERTAINLY there were *no insects* in the crop. The contents of this crop I presented to the Edinburgh Botanical Society, and they can now be seen in the Museum of the Society in the Botanic Gardens. In another crop, which I examined in April 1874, I found the contents to be *wholly the young shoots, leaves, and buds of larch.* I counted the extraordinary number of 918 buds alone in this 'crop,' besides the bits of shoots and leaves, which formed by far the *bulkiest* part of the whole. There were a few bits (three) of silvery lichen amongst the contents, but NOTHING ELSE.[1] These are given as *fair samples* of many crops I have examined, received chiefly from Perthshire, Mr. Brown having sent me about a score from Perth in 1874. In *none* of them did I ever meet with a *pine beetle*, or any other insect that would lead me to suppose that the bird preys upon insects, or had a preference for shoots which were infected by them. In fact, I should maintain that the bird prefers *clean, healthy, fresh* food, and has no taste for *damaged* or *decaying* vegetation of any kind. I have never examined the crop of a young bird taken out of the nest; but I have analysed the crops of several birds of the same year in July and August, and failed in every instance, to find any insects, so that, although I am aware that it is said in *books* that they are 'fond of insects, especially when young,' I am unable to corroborate the assertion. The nature and habits of the bird do not in any way lead me even to *suppose* it feeds on insects; but in other parts of the world —in Norway for instance—it may feed on different matter to what it does in Scotland.

"Since I made my investigations anent the injury done by the Capercaillie, etc., to forest trees, I have also investigated the injury done by *insects.* The injury done by the pine beetle to the Scots fir is in no ways analogous. The beetle does its injury *internally*, by eating the pith of the shoots and heart of the buds; the Capercaillie 'lops' the shoots, buds, and leaves *clean off*, and the one cannot by any possibility be mistaken for the other; besides, the injury is

[1] The pieces of lichen no doubt were picked up along with the other contents of the crop, and do not form a part of the regular food of the bird.—[J. A. H. B.]

done by the beetle in the middle of summer, and the most serious injury is done by the bird in the winter, when the beetles are hybernating in, or on, dead wood on the ground.

"The beetle attacks almost any pine tree, sick or healthy, any size or any age; only, as it climbs from the ground to the branches, small trees, say under 25 years of age, are those commonly attacked. Capercaillies, on the other hand, attack only *healthy* trees of any size or age, and will, in some instances, *return* to *the same tree for days continuously*, till it is completely stripped of its buds or growing points, and, of course, most seriously injuring it, and rendering it perfectly useless for timber. If the bird is kept within due limits (in numbers), the injury they do is immaterial to the general welfare of our forests; but if they become very numerous, they will certainly play havoc with the pine and larch plantations in their neighbourhood, especially young plantations. Black game, at certain seasons, are just about as injurious to young pine and larch trees." The above remarks by Mr. Dunn were evoked by special queries which I put to him regarding the possibility existing of the buds affected by the pine beetle being those chosen by the Capercaillie. As has been seen, Mr. Dunn is of the opinion that the bird prefers "clean, healthy, fresh food," and has "no taste for damaged or decaying vegetation of any kind." He does not consider that insects of any kind form part of their food, but he at the same time states that he "never examined the crop of a young bird taken out of the nest."

In this connection Mr. D. Brown, formerly naturalist for many years in Perth, writes that he has "stuffed very nearly 700 Capercaillies during the last twenty-two years," and "not in one single instance have I observed any form of insect in the very many crops I have emptied for inflation, or which may have burst during the operation of skinning the birds." Mr. D. Brown's and Mr. Dunn's evidence, however, must only be considered as negative, in opposition to the positive recorded testimony of a number of other naturalists. Mr. Robert Collett of Christiania distinctly states that the young birds "are at first fed almost exclusively on small soft-bodied insects" (‘*Birds of Europe*,’ by Dresser, Part xxi.). Mr. Charles Ottley Groom Napier, who has specially devoted his attention to compiling an account of ‘*The food, use, and beauty of British Birds*,’ gives as the food of the Capercaillie in May, "insects, leaves," etc. (p. 69). Various other authors of repute state the same, such as Lloyd (‘*Game Birds of Sweden*,’ pp. 4, 5). Indeed, I should myself be more surprised to learn that they did *not* eat insects, as all game birds and gallinaceous birds feed more or less upon insects and larvæ, especially when young; but I have not on that account neglected specially to make observations and inquiries in this direction, as the subject of the food of the Capercaillie is in itself of interest, all the more so when we find that a difference of opinion is expressed, as we find above.

Mr. Robert Collett and Professor Rasch of Norway likewise have never found insects in the crops of many Capercaillies they have dissected "in the season from autumn to spring." They are protected by law all the summer time. Mr. Collett adds that he believes they feed upon larvæ. "I cannot understand why they are often seen in numbers in places in our forests *where only* the insect larvæ are devouring the leaves of the birches or the blaeberry plants, if they do not feed upon these larvæ. But, unfortunately, no specimen must be killed in the season when these larvæ are present."

Herr Dr. Meves, of Stockholm, writes to me:—"I believe that these birds (Capercaillies) sometimes do as much good as harm to forests. Travelling in Jemtland I found young wood grouse, from three to ten days' old, having their throats and crops filled with *Tenthredo* or *Nemitus* larvæ, which are found on pine trees. A couple of wood grouse shot in

Dalerne about a month ago (his letter is dated 16th March 1878) had pine-needles (*Pinus sylvestris*), juniper-berries (*Juniperus*), and cranberries (*Vaccin. vitis-idœa*) in their crops."

Here is one little gleam of sunshine. They must destroy vast numbers of *Tenthredo* and *Nemitus* larvæ, even during the *at-present-supposed short period of their insect-feeding age.* But a question of interest is:—How long does this insect-feeding age continue? And another question of interest is:—How much good do they do in that time against the harm they do at others? These are questions the solution of which, I think, cannot fail to prove of economic interest to proprietors of forests, and which, if thoroughly investigated, may more than likely induce many to modify their wholesale condemnation of the Capercaillie. The rook, against which species Acts of the old Scottish Parliament were passed, now increases, and even with many farmers lives upon altogether a different footing than formerly. Many farmers know now that it would be a dangerous experiment to rid the country of rooks altogether. There are two sides to every question, and I have strong faith that if all the good the Capercaillie does were as thoroughly investigated as the evil has been, there would not be quite such a hue and cry against it.

A correspondent in Perthshire writes as follows:—"From the first of November up to the end of May the Capercaillie lives principally on Scotch fir 'sprigs.' Then, from the first of June to the end of October, he lives greatly upon insects, digs deep into ants' mounds[1] in search of food, and strips the bark off rotten trees in search of worms and beetles." Fancy hundreds, yes and thousands, of Capercaillies thus employed. Do they do no good?

I have quoted Mr. Dunn's letter fully, as it is perhaps the letter, amongst many others I have received, which places the

[1] Lloyd directs that ants' eggs be provided for the young birds when rearing them by hand (*op. cit.* p. 32).

damage done in the strongest light. I am still of opinion, however, that the final *results* exhibited in certain young trees are just as likely to have been caused by beetles as by the Capercaillies. If the buds are destroyed, whether by beetles *internally* or by Capercaillies, whether in summer or in winter, I believe the *results* will be the same, viz., as will be seen farther on—the trees becoming bushy, branchy, and stunted.

My friend Mr. J. J. Dalgleish—owning an estate in Perthshire, and having between 400 and 500 acres of pine, larch, and spruce plantations of different ages inhabited by Capercaillies—informed me that old trees are not so much damaged, because the birds cannot reach the terminal shoots. Whenever the trees get old enough, and the birds cannot any longer reach the top bud, the trees are safe. But on wood of a certain age they and black game together inflict serious injury. The result, in his opinion, is, that the trees become stunted and bushy, and unless a new upward growth takes place, soon become useless.

Becoming interested in the subject of destruction to forest plants and trees, I visited, in January 1878, a piece of ground of fourteen acres in extent, or thereabouts, situated in the midst of old pine wood of different ages, and which had been planted six years previously with pine seedlings on the above-mentioned estate. In one corner, facing the sun and the south, and protected on the north and west by older growth, the damage which the young trees had suffered was perhaps most apparent. Upon this estate Capercaillies are tolerably abundant, as many as four having been shot in one day by a party in 1877, and I have myself estimated the numbers seen in one day at least sixteen. Black game are scarce, and have been so for a number of years; but I have seen black game, driven out of the adjoining covers, and once rising out of the above enclosure. The stunted, bushy, or tufty young trees were pointed out to me, and I clearly saw where the central buds of the "leaders" had been picked out—or had dropped off!—resulting in side shoots taking the place of the leaders, and thus deforming the trees. This was affirmed to be the damage done solely by the Capercaillie, which birds were stated to stand upon the ground and pick out the buds from the leading shoots, when the trees were perhaps four or five years old, and had been planted perhaps three years. After reaching this age the trees are considered safe, as they are too high for the birds to reach the terminal shoot from the ground, and the young upper branches and leader are too young and tender to bear the weight of the bird. Becoming older still, the terminal shoot lengthens beyond the reach of the Capercaillie when it stands upon the upper branches, and the birds are then obliged to feed upon the side shoots. It was further stated that the trees, in consequence of this treatment when young, were delayed in their growth, and that many were destroyed and rendered useless.

Notwithstanding all this apparently strong evidence against the Capercaillie, I reserved the right of agent for the defence, with the hope of being able to pick holes still in the evidence; or, at all events, to mitigate his sentence. I saw a tree on the same property at least forty years old (on the authority of the proprietor), which was standing close to the young plantation, and with the same exposure, which was stunted in exactly the same way. It is needless to point out that that deformity could scarcely have been caused by Capercaillies, unless they attack old trees also, as Capercaillies were not introduced successfully into that district until 1864, when the trees would be thirty years of age; or, taking the imperfect introduction resulting in hybrids in 1856, and supposing the few birds there at that time capable of doing any damage, which is too unlikely, still the trees would be twenty-two years of age.

Mr. J. B. Hamilton, of Leny, amongst others, informs me that he has seen abundant evidence of their work and mischief, and adds :—" on Scotch firs of twenty to thirty years old. My own impression is that they are not so injurious to young Scotch fir plants of a foot or two in height as black game, which are very fond of the leading bud in these, and are one of the causes of that tufty appearance in these that you allude to."

There seems to be considerable diversity of opinion amongst those inimical to the Capercaillie as to the ages of the wood attacked. On the other hand, I have the following opinion from Sir Robert Menzies, which I quote in full :—" Neither do I blame them for being destructive to the plantations, a fault that is frequently found with them, and in consequence of which they are shot down. So far as my experience goes, they do not injure newly planted young trees in the same way that black game do, as they will not go to a plantation where there are no trees they can roost upon; they will not pick out the tops while standing on the ground,[1] and it is not till the trees in a plantation are somewhat grown that the Capercaillies settle in it, and then, as they are heavy birds, it is only the side shoots they can get at, as the top shoot is not strong enough to carry them. I am of opinion that it is no fault of the Capercaillies that plantations of Scotch fir are found not to do well, but that a very bad sort of that tree is now sold out of the nurseries that will never become trees; plants that do not make a leader like the true Scotch fir, but, on the contrary, have no stem, and abound in branches, both above and below. This sort of Scotch fir is an importation from the Continent, now of some years' standing, and young plants are now sold as 'true natives' that are seedlings taken from trees originally grown from this imported seed. When

[1] This opinion from Sir Robert Menzies was received quite independently of any suggestions of mine, or mention of other people's remarks.

these are not found to be doing well the Capercaillie is blamed; but I think it is the seedsman who is at fault, and more care should be taken to see that the young Scotch fir plants are of the right sort, when the plantation will be found to do well enough, though there are a good many Capercaillies in it."—(*Sir R. Menzies, Bart., in lit.*)

Further, Mr. Robert Collett of Christiania, in reply to inquiries I made of him regarding destruction done to forests in Norway and Sweden, writes:—"In Norway there is not any trace of destruction to the forests done by the Capercaillie. Certainly they do live in winter almost exclusively on the leaves of the fir (*Pinus sylvestris*), but they only take some shoots here and some there; for the most part from *old—or at least not young—trees.*

We find here (as well as in much other correspondence I have had on the subject) great discrepancy between the accounts of different authorities as regards the age of the wood at the time it is affirmed to be destroyed by Capercaillies. One affirms that it is only the young plants that are injured, and that the birds stand on the ground and pick out the "sprits," or leading buds; another states his conviction to be that they "will not pick out the tops while standing on the ground," and that they cannot reach the top shoots when the tree is grown, owing to these top shoots not being strong enough to bear their weight.[1] I cannot say if it is positively correct to state that they will not feed while standing on the ground; but it has been shown or recorded that they *do* sit upon the topmost shoots of smaller trees at "lek" time, in such localities as are destitute of larger trees. Dresser tells us ('*Birds of Europe*,' part xxi.), "They used some of the smaller trees for their 'lek;' and it was easy to dis-

[1] If they cannot reach the top shoots, then they must be content with side shoots, and thus become, to a certain extent, nature's pruners, as shown by the opinion of several naturalists.

tinguish those which had been used for that purpose, as the tops were bent down by the weight of the birds." If the bird does so at "lek," there is a probability that it will also do so at feeding-time. The general impression seems to be that it is only older wood that is attacked, and quite a number of my correspondents seem to be of this opinion, the damage done to young plants being not infrequently put to the charge of black game.

The editor of the Swedish hunting journal, '*Nya Yagore Forbundets Tidskrift*,' Stockholm, expresses his opinion in a letter to Herr Dr. Meves, that "there is certainly a local damage observed and complained of" in Sweden, "viz., in young plantations, where the plants are a foot or two in height. The havoc then done by the old male Capercaillie is sometimes considerable." Dr. Meves writes to me that his son—"The Inspector of Forests"—told him that there "is *very rarely* any complaint about damage done to young plantations by wood grouse."

I read and studied a little book '*The Larch Disease*,' by Charles MacIntosh (Edinburgh, 1860), with a view to obtaining parallel hints as to the natural history of the Scots fir. Reading this book rather strengthened my suspicions, and the statements of several correspondents (*op. cit.* p. 15, *et seqq.*) regarding imported seed, unhealthy seedlings, southerly exposures, hot suns, and late frosts, having something to do with the stunted appearance of some plantations. Accordingly, I wrote to Mr. John Hancock, of Newcastle, who being both a good ornithologist and experienced arboriculturist, would be likely to afford me good and unbiassed information.

The facts I have gathered from a correspondence with Mr. Hancock, and later correspondence, are as follows:—

There *is* a stunted form of fir supplied by nurserymen, which never makes a tree, but remains a dwarf, and bears cones. Mr. Hancock considers that these plants are generally reared from slips and not from seed. It has been the fashion of late years to plant this for cover for game.

Mr. Hancock considers the work of the Capercaillie as a natural state of pruning, which, in forests or woods of considerable extent, would result in little or no harm to the trees. In this Mr. Collett of Christiania would appear to agree in his remarks quoted above; and, indeed, I think all who consider at all that the balance of nature should be respected will have a similar opinion. If the population is too great for the extent of wood, no doubt damage must ensue; but, as pointed out by several correspondents, not to speak of my own inexperienced observations, this balance of nature could be preserved, or at least the evil remedied, if—as Mr. Hancock says—"all landed proprietors would plant two trees for every one cut down;" then the food supply would soon be in excess of the requirements of the birds." I have elsewhere shown that an increase of the population beyond what the balance of nature can stand results in the emigration of the surplus birds, and that an over-stock of birds cannot last for any length of time; a natural law proved by the very fact of the distribution and extension of range of species. If artificial feeding, or great excess of young wood be planted, the birds will increase in number accordingly; but this, too, will have its limit. The whole question appears to me to resolve itself into the question of the balance of nature and natural laws. If man, on the one side, infringes these laws by making two trees grow where nature only grew one, he must expect to see perhaps double the number of Capercaillies, because double the amount of food for them is supplied. But, on the other hand, in many parts of Scotland, the balance of nature has been disturbed by reckless cutting of forests *without replanting*. The consequence here is that the birds become proportionately scarcer, after no doubt doing a certain damage to the remaining trees. Man shoots them down on all occasions, forgetting or ignoring the fact that he himself was the first disturber of nature's laws.

Mr. Hancock considers that the stunted and bushy trees which I described to him will prove, in all probability, to be the work of a small beetle or weevil (before mentioned) which eats into the buds. Its work cannot be confounded (as already shown by Mr. Dunn) with what the Capercaillies do: "The latter cut the bud quite off, whilst the former leaves the dead buds attached to the stems."

"There seems to be no doubt but that the usual plantation plant is decidedly inferior to the native, and it is supposed that the seed is not pure—according to some authorities I have consulted—having been taken from plantation-trees which are not true Scots fir, but hybrids between the latter and some other of the family, or else that it is foreign seed of an inferior strain.

"It is certain that the Scots fir of modern plantations is an inferior plant. How much this may be due to uncongenial soil and injudicious treatment, not being an adept in forestry, I know not." (Dr. Buchanan White *in lit.*)

Although Capercaillies are said by Mr. Dunn only to attack healthy trees, I am not sure that *beetles* of the Hylurgus tribes may not be induced to confine their operations to the *inferior strain* of trees above mentioned. It is notorious that many parasitical insects are more apt to attack bodies which are in an unhealthy state than those which are healthy, and we have no stronger proof of this than in the presence of certain parasites in the human body.

Whether these insects attack all trees alike, healthy and unhealthy, native or imported, sound strain or unsound strain, indigenous old Scotch pine, as at Rannoch, or the younger growths, is more a question for the arboriculturist and entomologist than for an ornithologist, though other questions hinging upon the answers, from an ornithologist's point of view, *may*, or may not, prove of value.

Now, although the actual *work* of the beetle cannot be confounded with what the Capercaillies do, is it not quite possible that in many cases—I will not say in all—the *results*—viz. the stunting and deforming of the young trees—may have often been caused by the beetle, and the blame have been laid on the Capercaillie?

But before quitting this part of our subject—in which, let me say, there remains still a vast deal of investigation before we can consider it exhausted—I would shortly draw the attention of my readers to some other facts which must be considered in the connection of "damage done to forests."

In the course of an examination into the "damage done to forests" by another animal—the squirrel—at the distribution and increase of which, in Scotland, I am also working, I find that a favourite food of the squirrel, amongst a very extensive selection, consists of the shoots of *fir*, *larch*, and a variety of other trees. It seems to me difficult to account for the wide differences of opinion regarding the age of the trees attacked by the Capercaillie, otherwise than by supposing that there must be inaccuracy of observation either upon the one hand or upon the other, or by supposing that both sides of the question are more or less right or more or less wrong. I do not wish to cast a deeper shade over the already disreputable character which the squirrel bears amongst the foresters and woodmen of Scotland—its burden is perhaps already greater than it can bear—but I think it is quite possible, indeed more than likely, that the difference of opinion already taken notice of is traceable to the more regular attacks of the nimble little quadruped, whose weight is not so great as that of the Capercaillie, and would not prevent it from reaching the leading shoots at almost any age of the trees. Many will say at once that the traces of squirrel's work are visible upon the ground under the trees attacked, and can be at once identified, but I doubt if this is *always* the case. I fancy it would not *always* be easy to pick up the rejected husks cast down by the squirrel amongst undergrowth, or the long heather amongst which young fir seedlings are often planted. In the West Highlands, in localities where squirrels *are* not, how often do we find the ground underneath the birch-trees strewed with the fresh green shoots, and under the hazels with the "nut-bobbins;" and have we not watched the black game busy picking them off and letting them drop? Many a West Highland road by the margin of a lake or arm of the sea, at certain seasons of the year, are thus thickly strewn, but there are no squirrels there.

It is true the Capercaillie swallows the buds whole, and differs in its form of food thus far from the squirrel, but I think it unfair to the bird to put all the blame upon it for the "stunting" and "bushing" of the trees, old or young, without taking into consideration the other causes of the damage done. How much damage, done really by the squirrel, is laid to the charge of the Capercaillie? How much easier is it to see a Capercaillie in a young fir plantation, in which the plants are a foot or two in height, than to see the squirrel! The Capercaillie rises a hundred yards off, and seeks shelter in the higher woods, thus proclaiming his presence. The squirrel may, or may not, make for the higher woods, but at all events he more easily evades detection.

Yet another cause of damage to fir-plantations has been pointed out lately by Mr. J. Hardy, viz.:—the massing of wood-pigeons upon the top-shoots of young spruces and firs (*in lit.*); but I merely indicate these here to show that a much more exhaustive inquiry into the causes of damage to forests is necessary, before the amount due to each agent in it can be assigned with justice and impartiality.

Finally, on the subject of food. According to authors and correspondents:—In summer the food of the Capercaillie consists of various plants, fern-shoots, and buds of trees, such as alder, birch, and hazel, and acorns, where procurable; almost all sorts of berries, such as cranberries, cowberries, blaeberries, wild strawberries and raspberries, juniper-berries, and of insects. It also feeds on the leaves of the larch and Scotch fir, and sparingly of the spruce. A correspondent states that from 1st of November to end of May, or thereabouts, they feed upon vegetable food, and principally on fir-shoots; but in summer—*i.e.* from beginning of June to end of October—"they live greatly upon insects, digging deep into ants' mounds in search of food, and stripping off the bark of rotten prostrate trees, looking for worms or beetles."

So also do its congeners black game and grouse. "During spring and summer the black cock, as Mr. Lloyd informs us, feeds upon birch buds, tender leaves . . . berries, etc., and on insects and larvæ."

I will be glad if what I have said above will lead to a more thorough investigation of the *statistics* of damages done to forests, 1*st*, solely by Capercaillie; 2*d*, solely by squirrels; 3*d*, solely by insects; 4*th*, solely by wood pigeons; and 5*th*, solely by black game. Each of these subjects might well deserve separate and exhaustive treatment, but in such an investigation it is needless to say every separate act of destruction or damage should be distinctly and clearly brought home to *one* of the agents, and every side of the question be critically examined. Until actual, positive, unquestionable, and distinct evidence, and a large mass of carefully collected statistics be brought together, and viewed from every possible aspect, we cannot, I consider, with justice assign the amount of damage done to any one of these agents in particular.

Deeply aware of the imperfections of this part of the subject, I still hope it may lead to a more perfect treatment of it.[1]

[1] Since the above was written I have received a report upon the contents of four crops of Capercaillies from Dr. Buchanan White, but as these are, except in one case, unaccompanied by dates, they are not of much service. Three contained fir-needles, and a few buds and small stones. The fourth (7th October 1878) "a very little reddish grumous matter." As already seen, we want reports upon crops of birds *killed in summer*, and both of old birds and young, and without dates these reports are next to useless.

CHAPTER XXV.

DAMAGE TO GRAIN.

As to the damage done to grain, it certainly appears to be considerable where they frequent standing fields of corn. In the Crieff and Comrie district I have it from a reliable eye-witness that numbers of these birds alight in standing corn-fields, beating down the stalks with their wings each time they alight, and doing incalculable damage. On the other hand, in almost all other districts from which I have data, I am told that they are either "seldom if ever," or "never," seen in corn-fields. I am bound to say, however, that these latter reports do not emanate in most cases from parties who might take an impartial view of the matter, or from agriculturists who might not be blind to their shortcomings; and upon this point I still require data before arriving at conclusions. Meanwhile, I think it is safe to consider that any such damage is extremely local, and not general. A correspondent in Glenalmond informs me that the taste for grain has only lately been developed there. He says they are now *beginning* to feed upon the stooks on Balgowan estate (1877), and he adds with natural gladness, being a person interested in their welfare, "It will then be easier to rear them."[1]

Never have we heard that Capercaillies in a wild native

[1] The gamekeeper on Balgowan, however, has never seen them on arable ground, nor has he ever found grain in their crops.

state—as in Norway, Sweden, or Russia—have ever done damage to crops at any time, and I am borne out in this by several letters from Norwegian and Swedish naturalists.

Naturally the Capercaillie is a tame bird, at least in Scotland, since the restoration of the species, but where much disturbed or shot at, they soon learn to take care of themselves, and do so very effectually on the whole. In the nesting season they become often very tame, allowing passers-by to stand still and inspect them at a distance of a few paces.[1] It is usually, however, most difficult in an ordinary day's walk through the old fir woods to obtain a shot at them, and it is by driving the covers that they are usually killed in greatest numbers. I am not aware that the method—in practice in Norway and the north of Europe—of hunting them with a dog, and approaching after the birds are "treed," is much in practice anywhere in Scotland, although occasionally sportsmen stalk them with pea-rifle in hand. Mr. R. Anderson, of Dunkeld, gives me an amusing account of the squirrel-like actions of a hen Capercaillie when caught napping in a thick fir, dodging round the tree stem, and preferring this method to that of escaping by flight. It is under such circumstances sufficiently exciting and interesting sport, and worthy of a good marksman.

[1] On Sauchie property, in Stirlingshire, a cock bird—one of the first that arrived there—used boldly to attack any girl or woman passing along the avenue, close to which, and within 100 yards of the house, two hen birds were at the time sitting on eggs, and several times allowed himself to get caught.

CHAPTER XXVI.

CONCLUSION.

I HAVE now brought together all the information I have been able to collect upon these points in the history of the Capercaillie, which presented themselves to my mind as most worthy of attention in this country. I am, I trust, aware of my own imperfections in arranging these materials, and planning the form in which to lay them before my readers. I am aware especially of the unfinished state of the last two chapters—perhaps *the* most interesting to the landed proprietor and forester—but as I waited long after the other portions were brought up to the date of my latest items of information for fuller and more minute data regarding the *destruction done to forests,* but without receiving any additional statistics, and the summer having passed without my having the opportunity of examining a single Capercaillie's crop, I decided upon placing my MS. in the printer's hands, leaving till a future opportunity the more thorough investigation and treatment of this part of the subject.

I have not entered into the subject of the habits of the species for two reasons. First, *that* has already been most amply treated of by abler pens than mine; and second, having had comparatively few opportunities of narrowly watching the species, I am unable to add anything of sufficient novelty to this portion of its history, and I feel that were I to extend this essay to any further treatment than I have already done, I would be entering upon the dangerous habit called book-making.

I therefore conclude, hoping that some interest may be found in the foregoing pages; and if the information given is not new, at least it is carefully sifted, and, I believe, in all the more important passages, will be found to be reliable.

I shall be glad to receive further notes on the advance of the species, from those who may be sufficiently interested in the subject, in order to keep on record its further history; and, as already mentioned, I will be glad if the subjects treated of in the last chapter attract interest and further discussion.

APPENDIX.

DERIVATION OF 'CAPERCAILLIE.'

Vide CHAP. I. *Addition to Footnote at page* 3.

Since the first chapter of this Essay passed through the press, I have come upon the following passage in Joyce's '*Irish Names of Places*,' which appears to me to have peculiar significance in the connection. In his chapter upon Animals [chap. vii. p. 452, third edition, 1871], after telling us that "the transfer of a name from one species of animals or plants to another is a curious phenomenon, and not unfrequently met with," [*i. e.* in the Erse]—*op. cit.* p. 456,—he shows further on how this comes to be exemplified under *The Goat* and *The Horse.* He says "The word *gabhar* (gower), a goat, is common to the Celtic, Latin, and Teutonic languages: the old Irish form is *gabar*, which corresponds with Welsh, *gafar;* Corn., *gavar;* Latin, *CAPER;* Ang.-Sax., *haefer.* . . . The word *gabar*, according to the best authorities, was anciently applied to a horse as well as to a goat. In Cormac's '*Glossary*' it is stated that *gabur* is a goat, and *gobur* a horse. Colgan remarks that *gabhur* is an ancient Irish and British word for a horse," &c. (p. 459).

ANTIQUITY OF THE SPECIES.

Vide CHAP. III. page 14.

Professor Newton informs me that remains of Capercaillie *have* been found in a Roman layer at Settle, but up to date of going to press I have not learnt anything further than the statement.

FORMER HISTORY OF THE CAPERCAILLIE.

Vide CHAP. IV. page 15.

We have evidence of the prior history of the Capercaillie in the fact, that it was recognised by name by the ancient Britains, whose name for it was *Ceiliog Coed.* (*v.* '*Eng. Cyclop.*')

ADDENDUM TO CHAP. IV. page 33.

Longfield, in his Treatise on the '*Game Laws in Ireland*,' says that the "Wild Turkeys" of Act George III. must have been Capercaillies, and adds that they were not extinct so late as 1787. The earlier authors seem to have been sorely exercised as to what was the correct place the Capercaillie should take in the CLASS AVES. Thus, while, as we have seen, it received the names of 'pekokes' and 'Pavones sylvestris' and 'Wild Turkey,' Merrett, in his '*Pinax*' (1667, p. 179), puts "the Capricalca, Capricalze Scotis," among the *Aquaticæ Palmipedes*, and Charleton places it still further away in the company of the members of the genus *Anser.*[1]

RESTORATION IN IRELAND.

Vide CHAP. V. page 51.

Col. Edward H. Cooper of Markree Castle, Co. Sligo, has made attempts to introduce Capercaillies into that county. He writes—16th Feb. 1879—"My experience is so far not very hopeful. Three years ago I turned out a hen bird. She was not seen again. The following spring I got a sitting of eggs from Scotland; four hatched out, and lived for about six weeks; but I believe it is impossible to rear them by hand. This last autumn and winter I have turned out seven birds, but only one

[1] [*Vide :—ONOMASTICON ZOICON*, Plerorumque ANIMALIUM *Differentias & Nomina* Propria pluribus Linguis exponens. Cui accedunt *MANTISSA ANATOMICA ;* et quædam De Variis *FOSSILIUM* generibus. Autore *Gualtero Charletono*, M.D. *CAROLI II.* Magna Britannia. Regis Medico ordinario, & Collegii Medicorum Londinensium Socio. *LONDINI*, apud *Jacobum Allestry* Regalis Societatis Typographum. MDCLXVIII. (4to.) p. 98.]

cock. A hen or hens are constantly seen about, but the cock has not been heard of, so I am afraid there is very little chance that they will breed this spring, unless they cross with a black-cock. I have also turned out a good many fine healthy black game this autumn, and several, I know, are still about here. I also put a sitting of eggs under a pheasant last spring: they all hatched out but one egg: none of the young birds were ever seen: there were some heavy storms at the time, so the young chicks may have perished." It is hoped and expected that Col. Cooper will eventually succeed in restoring the Capercaillie to Ireland.

EXTENSION IN INVERNESS.

Vide CHAP. XIII. page 93.

Mr. Peter—Lord Lovat's factor at Beauly—informs me that "one of the old native breed of Capercaillies was trapped at Struy (?), about 50 or 60 years ago. None have been seen since that time." This date would be placed at from 1819 to 1829. I wrote for further particulars and more exact dates, but have not received any up to the date of going to press.

Printed in Dunstable, United Kingdom

91343135R00105